MICROSCALE AND MINISCALE LABORATORY INVESTIGATIONS IN ORGANIC CHEMISTRY

Paul G. Johnson

and

Sujata Patil

DEPARTMENT OF CHEMISTRY AND BIOCHEMISTRY

DUQUESNE UNIVERSITY

2nd Edition

KENDALL/HUNT PUBLISHING COMPANY
4050 Westmark Drive   Dubuque, Iowa 52002

TABLE OF CONTENTS

PREFACE

The experiments in this laboratory manual were written to help students better understand the theoretical principles and concepts covered in the lecture portion of the two-semester sophomore organic chemistry course. There are experiments in the manual that examine the kinetic and mechanistic aspects of organic reactions as well as the synthetic methods available for preparing organic compounds. In lecture, students learn to predict reactivity from the structures of specific organic compounds. In lab they will verify these predictions. Also, using various chemical and spectroscopic methods, students will learn to differentiate compounds containing a variety of functional groups.

The experiments are especially designed for large lab sections of 25 students or more whose size prohibits the use of a fume hood to perform routine functions. This is accomplished via a microscale approach that minimizes student exposure to flammable and toxic organic compounds. There is also a cost-benefit advantage to microscale chemistry. The quantity of chemicals needed and the volume of waste generated decreases by 90% when microscale techniques are implemented. The use of microscale equipment also allows techniques such as filtration, distillation, and extraction to be performed more rapidly than their macroscale counterparts. However, this manual also contains several miniscale experiments that permit the student to develop an expertise in using macroscale glassware such as the separatory and Buchner funnels as well as standard ground glass equipment. We have found that sophomore organic chemistry students have adapted well to microscale synthesis. After a short period of time, they experience little or no difficulties in weighing, transferring, extracting, crystallizing, or distilling milligram quantities of organic compounds.

The manual is also designed for colleges or universities that use graduate students as instructors or instructional aides. Separate lab notebooks that are sometimes difficult to grade objectively are not required. All of the data, observations, calculations, results, and discussions are entered directly into the manual. All of the pages are perforated so that they can be easily removed and turned into a graduate student for grading. Unless noted otherwise, all of the experiments can be performed within a 3-hour period.

INTRODUCTION

I.   Safety in the Organic Chemistry Laboratory

Operating safely in the laboratory begins with having a complete understanding of the experiment you are performing. It is important that you thoroughly read and comprehend the introductory material and procedure for each experiment. This permits you to operate quickly, efficiently, and safely. Although most of the compounds you will handle are toxic or flammable, few are absorbed through the skin. It is just a matter of washing the affected area thoroughly upon skin contact. However, there are a few instances where skin contact or inhalation of vapors should be avoided. Although your lab instructor will caution you when these experiments are performed, you will also be required to look up and record the safety requirements and hazards involved in handling the compounds for each experiment. This information can be found in a variety of chemical catalogues and handbooks that are stored in the organic lab or stockroom. Another useful source is the *Merck Index* that you can easily find in the reference section of the library.

The best way to work safely in the lab is to treat each chemical with respect. In most cases your risk is minimized because you are using only milligram and milliliter quantities of material. It is important that you learn to weigh, measure, and transfer solids and liquids precisely without spilling any on yourself, your clothing, the balances, or the bench tops. This is just a matter of mastering proper techniques and taking your time. If you do spill any chemicals, you must follow appropriate clean-up procedures immediately. The material you spill as well as the chemicals and solutions you use in performing the experiment must be disposed of in appropriately labeled waste containers found in the fume hoods. All colleges and universities are required by law to develop guidelines for collecting and disposing of all chemical waste generated in the teaching laboratories. Also note that there are separate waste containers for such items as halogenated, nonhalogenated , solid organic, and aqueous waste. Contaminated paper towels and filter paper are also treated separately. It is an ethical (and legal) responsibility to see that your waste is placed in the correct container.

As you did in general chemistry, it is also important that you learn a set of rules for operating in the organic chemistry laboratory. They can be found on the following pages. You will be administered a safety quiz at the beginning of the fall semester just before you begin the first experiment. A 100% on the quiz is required.

SAFETY RULES FOR ORGANIC CHEMISTRY LAB

1. EYE PROTECTION MUST BE WORN IN THE LAB AT ALL TIMES. REMEMBER YOU WILL NOT GET A SECOND CHANCE TO GROW AN EYE.

   - Approved safety goggles must be worn by all students. They fit over ordinary eyeglasses.

   - Contact lenses should never be worn in the laboratory. Chemicals and vapors may infuse under the contacts and cause irreparable eye damage. Furthermore, they cannot be removed rapidly enough if a reagent accidentally splashes in the eye.

   - In case something splashes into your eye, wash thoroughly with water from the eyewash station, and seek immediate medical attention. Be sure you know the location of the eyewash bottle nearest your desk.

2. Fire is a special hazard in the chemistry laboratory. Regard all organic solvents as flammable and toxic. Be sure you know the location of the nearest safety shower, fire blanket, and fire extinguisher. In case of fire:

   - Extinguish all nearby burners and remove all combustible materials.

   - A small fire may be extinguished by covering it with an empty beaker that cuts off the oxygen supply to the fire.

   - Larger fires can be extinguished with a carbon dioxide fire extinguisher that should be aimed at the base of the fire. Do not use water.

   - **If your clothing catches on fire walk (don't run) to the nearest safety shower** and use it. Never use a fire extinguisher on a person.

   - Be sure no flammable solvents are near open flames.

   - No smoking, eating, or drinking is permitted in the laboratory

3. Wear proper clothing in the lab. Clothing is your first defense against chemicals. No shorts are permitted to be worn in lab.

   - Full-length pants are required for males and strongly recommended for females. Pants made of cotton or wool are preferable. Full length sleeves and a lab coat are recommended.

   - Shoes that fully enclose the foot are to be worn at all times. No sandals or bare feet are allowed. This protects you from spills or broken glass.

- Long hair should be tied back. The oil in the hair makes it very flammable. Be sure that neckties are tucked into the shirt.

- Coats and bookbags are not to be brought into the lab. Do not place any items in front of the doorways to the lab.

4. Chemicals and equipment are to be handled carefully. Be sure you are familiar with the properties and hazards of the chemicals before you come to lab.

- When diluting concentrated acids and bases, always add the acid or base to the water, never the reverse.

- Sulfuric, nitric, and hydrochloric acids are frequently encountered in the lab. They are very corrosive.

- When warming liquids in a beaker or test tube, be careful not to heat too rapidly. This might cause it to splatter on you or someone else. Never point a test tube toward anyone while heating.

- NEVER USE MOUTH SUCTION TO FILL PIPETTES.

- Be especially careful when using syringe needles. It is very easy to puncture your skin with a needle. Never use a dirty or rusty needle. Also note that there are special red, plastic containers for disposing of used needles.

- Beware of causes of electric shock. Do not use equipment with worn wiring. Do not handle electrical connections with damp hands.

- Never heat flammable liquids with a flame.

- There are two categories of flammable solvents:

    a) HIGHLY FLAMMABLE: examples are diethyl ether, pentane, hexane, and low boiling petroleum ether.

    b) VERY FLAMMABLE: examples are acetone, isopropyl alcohol, methanol, toluene, and tetrahydrofuran.

- Carefully read the labels on all chemicals before using.

- Make sure that the apparatus you are using is securely supported.

- To determine the odor of a chemical, waft the vapors gently toward your nose.

- Special care is needed when dealing with a mercury spill from a broken thermometer. Notify your lab instructors. They have a special device for collecting mercury.

- If you spill any chemicals on your skin, wash the affected areas thoroughly with water.

- A safety shower can be used for any large chemical spill on skin or clothing.

5. Beware of glass. It is the most common cause of cuts.
   - Use extreme care when inserting a glass rod or thermometer into a rubber stopper. Lubricate with glycerine, grasp with a towel, and slowly insert with a twisting motion.

   - Do not remove glass from a stopper. Instead, cut the stopper

   - Inspect all glassware for cracks and sharp edges before washing them at the sink.

   - Hot glass looks the same as cold glass. Carefully check the temperature of heated glass before picking it up

   - All broken glass should be placed in a labeled glass container. Do not place anything but glass in these containers.

6. WASTE DISPOSAL. Do not pour solutions down a drain unless you have been specifically instructed to do so. There is a specific section of a hood in the laboratory designated for waste. **Be sure to place the correct waste in each bottle**.

7. Wash your hands thoroughly with soap and water before leaving lab and again in the restroom before you leave the building. Never go from lab to lunch or dinner without washing your hands.

8. Proper conduct is required in the lab at all times.

9. REPORT ALL ACCIDENTS TO YOUR LAB INSTRUCTOR. The most common lab accidents are as follows:
   - Pulling steel wool or sponge rather than cutting it with a pair of scissors.
   - Puncturing the skin with a syringe needle
   - Grabbing a hot plate or piece of hot glass before it cools
   - Washing a flask or beaker at the sink and not knowing that it has sharp edges or is already broken.

10. If the fire alarm sounds, shut off all of your equipment. Leave the building as quickly and quietly as possible. Do not use the elevators.

II. Chemicals and Solvents

To navigate through organic chemistry lab it is important to know some of the properties of the compounds you will be using. These properties are very much dependent on the reactive group present on the molecule. This reactive moiety is referred to as a functional group. In fact, the pedagogical method of teaching organic chemistry in the lecture part of the course utilizes the functional group approach. Although the properties of functional groups such as aldehydes, ketones, carboxylic acids, and esters are not covered in lecture until second semester, it is necessary that you be able to recognize a few basic functional groups at the beginning of your laboratory experience. A list of these can be found in Experiment 2, "solubility properties of organic compounds."

A solvent is a liquid used to dissolve various compounds so that a reaction can take place. In organic lab, solvents can also be used in various purification and extraction methods as you will see later. The names and structures of the very important solvents we will be using are also listed in Experiment 2.

Solvents can be classified in a number of ways. Solvents are either **protic** or **aprotic**. The protic solvent is one that is a hydrogen-bond donor, that is, it possesses an N-H or O-H bond. An aprotic solvent does not have an N-H or O-H bond.

Some liquids are classified as donor solvents. These are molecules that can donate a lone pair and thus behave as Lewis bases. Diethyl ether and methanol are appropriate examples. You will see their importance later when we prepare a Grignard reagent. Nondonor solvents do not have available lone pairs. Examples in this category are hexane, benzene, and toluene.

Solvents are also classified as being **polar** or **nonpolar**. A polar solvent is one that effectively surrounds and solvates a cation or anion. This helps to shield one ion from another thus inhibiting their interactions with one another. As will be discussed in lecture, the polarity of a solvent depends on its dielectric constant (abbreviated with the Greek letter epsilon, ε) that is part of a mathematical equation based on an electrostatic law. Polar solvents have high dielectric constants. Solvents with an $\varepsilon > 15$ are considered to be polar. Water has an $\varepsilon = 78$, acetone 21, and hexane 2. A table with a list of important solvents in order of decreasing dielectric constants can be found on the next page.

In the organic laboratory it often necessary to classify solvents in other ways. It is important to know which ones are flammable or which ones are miscible with water. Knowing the boiling point of a solvent is often necessary. In many cases a solvent needs to be evaporated at the end of an experiment to isolate the product. The lower the boiling point of the solvent the more convenient the evaporation. For example, those with a boiling point under $80°C$ can be evaporated on a steam cone in the fume hood. These properties are also listed in the table on the next page.

TABLE OF SOLVENTS IN DECREASING ORDER OF DIELECTRIC CONSTANTS

Solvent	Symbol	Protic	Donor	Boiling point (^{o}C)	Flamm.	Water Miscib.
Water	H_2O	Yes	Yes	100	No	
Dimethyl sulfoxide	DMSO	No	Yes	189	Yes	Yes
Acetonitrile	CH_3CN	No	Yes	82	Yes	Yes
Dimethylformamide	DMF	No	Yes	153	Yes	Yes
Nitromethane	NO_2CH_3	No	No	101	Yes	No
Methanol	CH_3OH	Yes	Yes	65	Yes	Yes
Ethanol	EtOH	Yes	Yes	78	Yes	Yes
Acetone		No	Yes	56	Yes	Yes
Methylene chloride	CH_2Cl_2	No	No	40	No	No
Tetrahydrofuran	THF	No	Yes	66	Yes	Yes
Acetic acid	HOAc	Yes	Yes	117	No	Yes
Ethyl acetate	EtOAc	No	Yes	77	Yes	No
Propyl acetate	n-prOAc	No	Yes	102	Yes	No
Chloroform	$CHCl_3$	No	No	61	No	No
Diethyl ether	Et_2O	No	Yes	35	Yes	No
Benzene	C_6H_6	No	No	80	Yes	No
Toluene		No	No	111	Yes	No
Carbon tetrachloride	CCl_4	No	No	77	No	No
Dioxane		No	Yes	101	Yes	Yes
Cyclohexane		No	No	81	Yes	No
Petroleum ether		No	No		Yes	No
Ligroin		No	No		Yes	No
Hexane		No	No	69	Yes	No

SECTION 1

Miniscale and Microscale Methods, Operations, and Techniques

The purpose of this section is to introduce you to the operations that are performed repetitively in the organic chemistry laboratory. You will learn to manipulate the equipment and master the techniques involved in the isolation and purification of organic compounds. You will learn to use the glassware in your microscale kit as well as some of the macroscale equipment in your lab drawer.

Normally, the synthetic organic chemist begins by reacting a starting material with a reagent to produce a desired product. Although the reaction is designed to maximize the yield of this product, a small amount of an undesired product, commonly referred to as the "**side product**," often forms. The final reaction mixture may also contain some of the starting material that didn't react. The desired product must be purified by removing the unreacted starting material and side product. This typically involves using common separation techniques such as extraction, distillation, or chromatography. You will learn these techniques in this section.

"starting material" + "reagent" →   "desired product" + "side product"
                                           + "unreacted starting material"

Once isolated, the purity of the product must be determined. This usually involves "taking a melting point" or "doing TLC." Both of these techniques are introduced in this section. Occasionally purity is ascertained from infrared spectroscopy. This technique is introduced early and can be found in the last section of the laboratory manual.

Operations in organic chemistry lab also involve recognizing functional groups and understanding how they affect the physical properties of the compounds in which they are found. It is also necessary to understand the properties of the various solvents that are commonly used. An introductory solubility experiment will help you understand these concepts.

EXPERIMENT 1

DETERMINATION OF MELTING POINTS

INTRODUCTION

Often the objective of a laboratory chemist is to identify an unknown crystalline compound and determine its purity. This characterization can be accomplished by determining a particular physical property of the compound. Recall from general chemistry that a **physical property** is one that can be determined without changing the compound into a different substance. Examples of physical properties are density, specific gravity, melting point, and boiling point. To an organic chemist the most useful physical constant is the melting point which is the temperature when a solid changes to a liquid. This is normally accomplished by putting a small amount of the material in a **melting point capillary tube** and placing the tube in a **melting point apparatus**. The apparatus is equipped with a thermometer, a heating block for raising the temperature slowly, and a magnification eyepiece for observing the melting of the small amount of material in the capillary tube. This can be used to determine the temperature at which the solid begins to melt (appears to turn "liquid') and the temperature at which the compound has completely melted. Together these comprise the "melting point range." Pure organic compounds typically have a small melting point range of 1 or 2 degrees.

This can be used to characterize a compound by simply comparing the measured melting point to the literature value. For example, a chemist has an unknown that she believes is cinnamic acid. Using the melting point apparatus she determines a melting point range of 132-133°C. Checking the Merck Index she discovers that the melting point of cinnamic acid is 133°C. This is a good indication that her unknown is cinnamic acid. However it is usual to do other tests to verify this conclusion. There are other compounds in nature that also melt at 133°C.

The melting point range can be used to determine purity. Note that the melting point range for the sample above is small. This suggests that the cinnamic acid sample is very pure. Impurities dissolved in a solid sample will cause the melting point to be **broadened** and **lowered**. Recall from general chemistry that melting point (same as freezing point) depression can be expressed by the equation:

$$\Delta T = K_f m$$

where K_f is the molal freezing point constant and m is the molality of the solution. Therefore a melting point range of 3 degrees or more suggests that the solid is impure. The broader the range the greater the amount of impurity present. For example, a pure sample of succinic acid melts at 190°-191°C. A mixture of 80% succinic acid and 20% urea exhibits a melting point range of 178°-186°C. Note, however, that melting point depression is a solution effect. The impurity must be dissolved in the solid compound.

Insoluble impurities such as several pieces of filter paper or a speck of charcoal in a solid sample will not depress the melting point.

The fact that soluble impurities broaden and lower melting points can be used to a chemist's advantage. Recall previously that a chemist determined a melting point of 132°-133° for a sample that she thinks is cinnamic acid. This conclusion can be verified by doing a **"mixed melting point."** An authentic sample of cinnamic acid obtained from the stockroom can be mixed with her sample. This must be done carefully by pulverizing and mixing the two very carefully in a mortar and pestle or on a watch glass to obtain a true solution. A melting point of this new mixture is then determined using the melting point apparatus. If her solid sample is indeed cinnamic acid, then a short-range melting point around 133°C should be obtained. If her sample is not cinnamic acid, then a mixture of two different compounds results and the melting point will have a broad range below 130°C.

Some organic compounds have a high vapor pressure sublime and before they melt. Sublimation is the process whereby a solid changes directly to the gaseous state without becoming a liquid first. This can be overcome by using the **"sealed capillary method."** After placing the sample in the capillary tube, the open end is sealed under vacuum. This prevents the sample from evaporating before it melts.

Why do solids melt? Crystalline solids contain molecules that are very highly ordered. The interactions that hold these molecules together are those that were typically discussed in your general chemistry course. In order of increasing strength these are (a) London Forces, (b) dipole-dipole interactions, and (c) hydrogen bonding. **Note that covalent bonds are not broken when a substance melts.** Upon heating a solid sample in a melting point apparatus, energy will be absorbed. The intermolecular bonds will begin to rotate and vibrate until a temperature is reached (**melting point**) where less ordered intermolecular interactions occur, and the solid becomes a liquid. The stronger the original intermolecular interaction, the higher the melting point. Because the overall number of interactions increase with size, the melting point of molecules tend to increase proportionally to molecular weight. Two constitutional isomers typically melt at different temperatures. The more symmetrical of the two usually melts at a higher temperature because its symmetry permits the molecule to form a more stable crystal lattice.

Procedure

1. Complete Table 1 in the data section before coming to lab. You need to look up the melting points in the Merck Index for the compounds listed. While there, also record the safety precautions listed in the index.

2. You will determine the melting point of an unknown assigned by your lab instructor. It will be one of the compounds listed in Table 1. Place a pea-sized amount of the compound on a clean watch glass. Press the open end of a melting point capillary tube into the unknown so that several millimeters of the powder enter. Then turn the capillary over and tap it against the bench top until the compound settles at the bottom of the closed end of the tube. A small amount will suffice because the apparatus you will be using has a magnifying glass.

3. Place the capillary tube (closed end down) in one of the wells of the melting point apparatus. Turn on the instrument and notice that a small light enables you to see your unknown through the eyepiece. Adjust the voltage until the temperature rises at the rate of about 2 degrees per minute. A faster rate will not enable the unknown to equilibrate with the temperature of the apparatus and your observed melting point will be in error. In Table 2 record the temperature range at which your unknown begins to melt and completely melts. Using the table of melting points, decide which compound is your unknown.

4. To verify this is indeed your unknown, you will do a mixed melting point. In the hood you will find a jar containing an authentic sample of your unknown. Place a pea-sized amount on a watch glass along with an identical amount of your assigned unknown. Using a stirring rod take about 2 or 3 minutes to thoroughly mix the two samples. Take a melting point of this mixture using steps 2 and 3 above and record the range in Table 2.

5. Clean and dry your watch glass and stirring rod and do another mixed melting point using your assigned unknown and a sample from the table whose melting point is second closest to your unknown. *Be sure you mix the two samples thoroughly.* Record this melting point range in Table 2.

Name _____ Sec. _____

DATA

Table 1. Melting Points

COMPOUND	m.p.	Safety / Hazards
Stearic acid		
Vanillin		
Benzoic acid		
Urea		
Salicylic acid		
D-tartaric acid		
Succinic acid		

Table 2. Unknowns

Assigned Unknown	m.p. =
Mixed melting point (step4)	m.p. =
Mixed melting point (step 5)	m.p. =

Name _____ Sec. _____

QUESTIONS

1. What is the identity of your unknown sample? Explain how you made this determination.

2. Discuss in detail how you used the mixed melting point method to verify the identity of your unknown.

3. Look up the properties of the solid compound, naphthalene, in the Merck Index. Based on your reading, what special precautions are needed if you were to determine the melting point of naphthalene?

4. Below are the structures of two isomeric hydrocarbons with their listed melting points. Even though they have the same molecular weights, the melting point of anthracene is nearly twice that of phenanthrene. Explain.

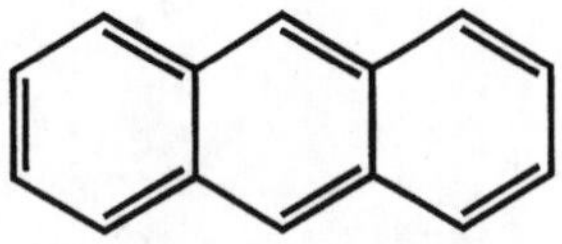

m.p. 218°

Anthracene

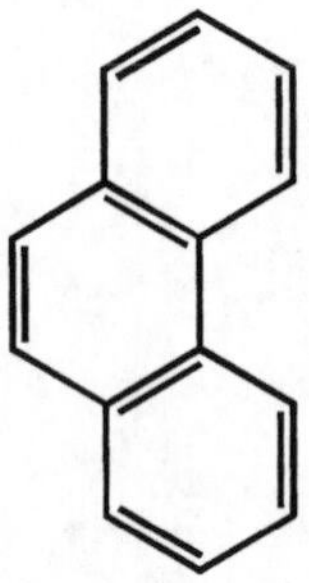

m.p. 100°

Phenanthrene

5. If the temperature of the melting point apparatus rises too fast, the solute in the melting point capillary will not have time to equilibrate with the temperature of the apparatus. Will this cause the reported melting point to be too high or too low?

EXPERIMENT 2

SOLUBILITY PROPERTIES OF ORGANIC COMPOUNDS

INTRODUCTION

Organic chemistry involves the study of the structure-activity relationships of organic compounds. In this experiment you will examine how the structure of various solutes and solvents affect a physical property called **solubility**. This can be the ability of a solid solute to dissolve in a liquid solvent producing a solution, or the ability of two solvents to homogeneously mix. Two solvents that do so are said to be **Miscible**. Two solvents that do not mix will appear cloudy and settle out to form two separate layers. These two solvents are said to be **Immiscible**.

To better understand them, chemists have organized organic compounds into categories called "**functional groups**." The more common ones are listed in this experiment. Organic compounds with the same functional group tend to have the same properties and reactivities. For example, compounds with the carboxylic acid functional group (--COOH) usually react with strong bases such as NaOH. The more alcohol (--OH) functional groups a molecule has, the more soluble in water it will be. And so on. We will begin to examine these relationships now and will continue to study them in subsequent experiments.

There is a general solubility rule that states: "**Like dissolves like**." What this essentially means is that solute molecules that are relatively nonpolar tend to dissolve in nonpolar solvents while polar solutes dissolve in polar solvents. Therefore, you need to learn how to assess the polarity of an organic compound based on its "R" group and its functional group. You have already learned some of the ways of doing this in the lecture part of the course. In organic lab, however, one can make several generalizations that often work.

1. The more carbon-hydrogen and carbon-carbon bonds present, the more nonpolar the molecule, and the more likely it will dissolve in hydrocarbon solvents such as ligroin, petroleum ether, or toluene. The less likely it will be soluble in the polar solvent, water.

2. Compounds with +/- charges, or hydrogen bonding donor/acceptor functional groups (OH or NH) will more likely dissolve in water or alcohol which also possess these functional groups.

On the next page you will find a list of the solutes that will be tested for their solubilities. Using the Merck Index, look up their structural formulas and draw the appropriate condensed structure under each name. *You must do this before you come to lab*. While there, look up the precautions for the safety/hazard column.

Name _____ Sec. _____

TABLE 1.

Compound	Structure	Safety/hazards
Stearic acid		
Glucose		
Benzophenone		
Benzoic acid		
Succinic acid		
Citric acid		
Salicylic acid		
Cholesterol		

Solubility is due to intermolecular interactions between the solute and the solvent.

+/- solutes dissolve in the polar solvent water due to "solvation" which permits separation of cation and anion.

For example, sodium acetate, $H_3C-\overset{\overset{O}{\|}}{C}-O^-\ Na^+$ is soluble in water

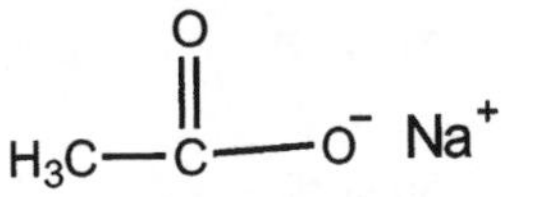

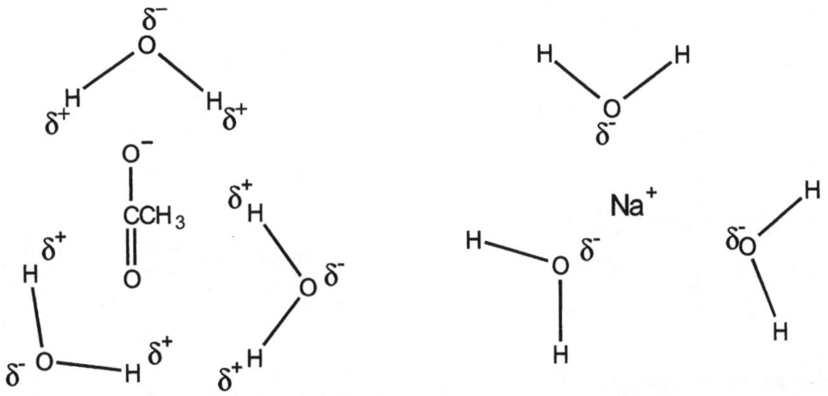

Acetic acid dissolves in water because the carboxylic acid functional group hydrogen bonds to water.

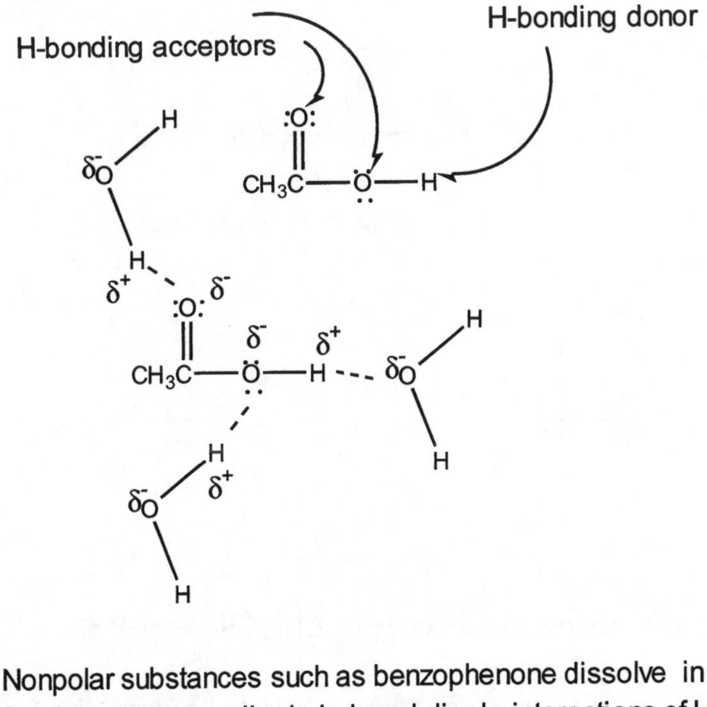

H-bonding acceptors

H-bonding donor

Nonpolar substances such as benzophenone dissolve in ligroin and pet ether because of the instantaneous dipole-induced dipole interactions of London dispersion forces (Van der Waals Forces).

COMMON SOLVENTS USED IN THE ORGANIC CHEMISTRY LAB

CH_3-OH

methanol

methyl alcohol

CH_3CH_2-OH

ethanol   EtOH

ethyl alcohol

$CH_3CH_2-O-CH_2CH_3$

diethyl ether

"ether" Et_2O

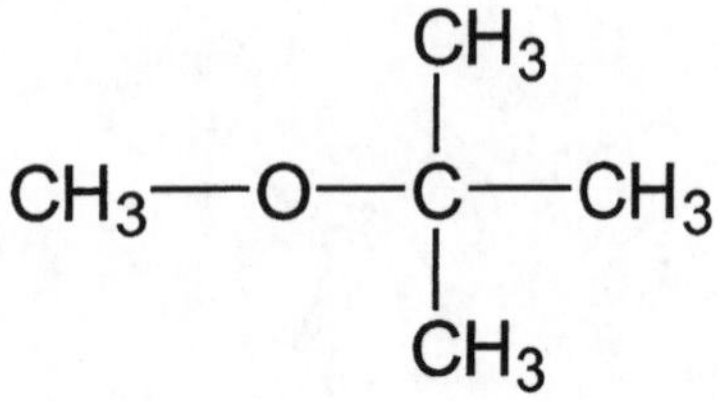

tert-butylmethyl ether   (TBME)

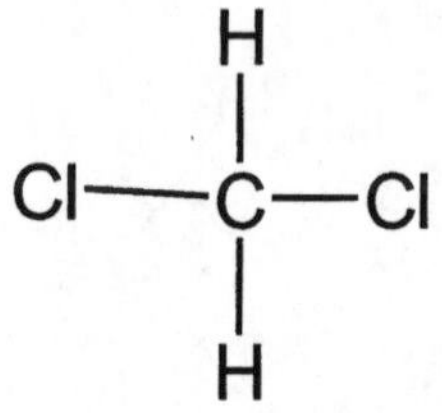

Dichloromethane

methylene chloride

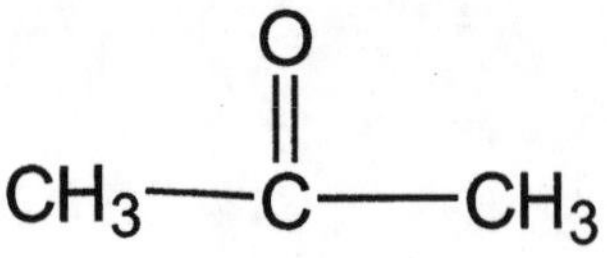

(CH_2Cl_2)

acetone

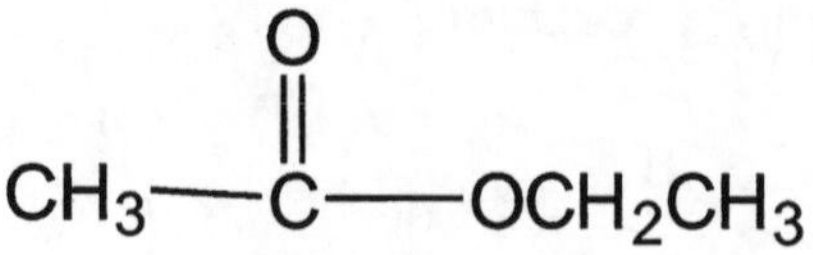

ethyl acetate (EtOAc)

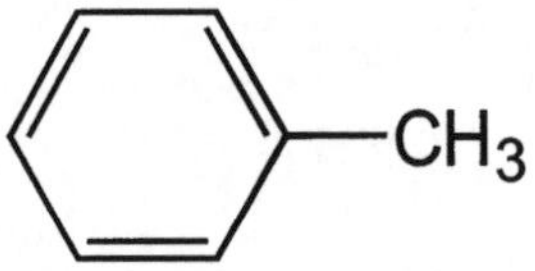

toluene

Petroleum ether: a mixture of hydrocarbons, chiefly pentane

Ligroin: a mixture of saturated hydrocarbons, chiefly hexane and cyclohexane

A FEW IMPORTANT FUNCTIONAL GROUPS

1. The double bond

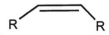

R   R

2. the alcohol

R—O—H   (ROH)

3. the ether

R—O—R   (ROR)

4. the amine

$$
\begin{array}{c}
H \\
| \\
R—N—H
\end{array}
$$ (RNH$_2$)

5 the amide

$$
\begin{array}{c}
O \\
\| \\
R—C—NH_2
\end{array}
$$

6. the aldehyde

$$
\begin{array}{c}
O \\
\| \\
R—C—H
\end{array}
$$ (RCHO)

7. the ketone

$$
\begin{array}{c}
O \\
\| \\
R—C—R
\end{array}
$$

8. the carboxylic acid

$$
\begin{array}{c}
O \\
\| \\
R—C—O—H
\end{array}
$$ (RCOOH)

9. the salt of a carboxylic acid

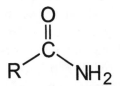

10. the ester

$$
\begin{array}{c}
O \\
\| \\
R—C—O—R'
\end{array}
$$

Name _____ Sec. _____

PROCEDURE

Part I

A.

In your microscale kit you will find calibrated, 5-ml test tubes. You need to remember that these are always referred to as "**reaction tubes**." Do not wash them. Go to the hood and add 1 ml of petroleum ether to each one. Then add 1 ml of water to the first tube, 1 ml of ethanol to the second, 1 ml of ethyl acetate to the third, and 1 ml of toluene to the fourth one. Mix each one by securely holding the top of the tube and flicking the bottom of the tube with your index finger. Note whether the four solvents are miscible or immiscible with petroleum ether by observing whether two layers form. Record your observations below. When finished, pour the contents of each tube into the part I.A. nonhalogenated waste container.

The same four reaction tubes are now going to be filled with water. From the first part of the experiment you should have observed that petroleum ether and water don't mix. This is an important concept to remember while in organic lab. You can't wash out a residual nonpolar organic solvent still in your reaction tubes with water. THEY DON'T MIX. To overcome this problem we use a universal solvent that mixes with both polar and nonpolar things. This solvent is ACETONE. That is why you will find it in squirt bottles throughout the lab. Squirt some acetone into your reaction tubes and pour it into the nonhalogenated organic waste container. Do this several more times. Then rinse the tubes with water several times. No problem now because acetone and water are miscible.

Name _____ Sec. _____

B.

Add 1 ml of water to each reaction tube and test its solubility with 1-ml portions of ethanol, methanol, ethyl acetate, and methylene chloride. Observe the miscibility and record. Note that the last tube must be disposed of in the halogenated waste container. Record your observations below.

Part II.

A.

Remove the 6 larger test tubes from your drawer and add 40 drops (about 2 ml) of DeI water to each one. Using a microspatula add a pea-sized amount of citric acid to one of the tubes. Mix and note whether the solute (citric acid) dissolves in the solvent (water). This may take several minutes and there may be a few specks that don't dissolve. Repeat this process with benzophenone, succinic acid, cholesterol, glucose, and stearic acid. Record all of your observations. Dispose of the contents of the tube into the aqueous waste container.

To do the next part, the 6 tubes must cleaned and refilled with methylene chloride. How are you going to do this? Look at the results from part I to see whether CH_2Cl_2 and H_2O mix.

B.

Once clean, fill the 6 test tubes with 2 ml of methylene chloride and test the solubility of benzophenone, citric acid, cholesterol, stearic acid, salicylic acid, and glucose in this solvent. Record your results below. Dispose of the contents in the halogenated waste container and rinse several times with acetone then water. The water washings can go down the drain.

Part III.

Place 5 ml of DeI water in one of the large test tubes. Then add a pea-sized amount of salicylic acid. Stir with a stirring rod and note that it doesn't dissolve. You are now going to heat the test tube in a sand bath on your lab bench. Your lab instructor will direct you on how to use one. You need to adjust the variac so that you can hold the test tube inside without burning your hands. Stir while heating for 5 or 10 minutes and record your observation. Then place the test tube in a beaker until the end of the period. Observe and record. What waste container should you use for part III?

Part IV.

Recall from general chemistry that acids naturally react with bases producing conjugate acids and bases. This process is called neutralization. Therefore, it should not be surprising that carboxylic acids (functional group #8) react with strong bases such as sodium hydroxide:

equation 1

$$RCOOH \quad + \quad NaOH \text{ -------->} \quad H_2O \quad + \quad RCOO^- Na^+$$

   acid   base   conjugate acid   conjugate base

Even though the carboxylic acid itself may not dissolve in water, upon neutralization it forms the salt of a carboxylic acid (functional group #9), which should be soluble in water because of the very polar +/- charges.

Note that NaOH and HCl are both corrosive. Handle carefully using dropper bottles.

Place a pea-sized amount of benzoic acid in each of two large test tubes. To one add about 1 ml of water and note whether the benzoic acid dissolves. Then add about 1 ml of 5% NaOH to the other test tube and mix. Record your observations. You should be able to rationalize the results with respect to what was just discussed above.

Then to the test tube you just added the base to, add about 2 ml of 3M HCl and note the results.

Record your results below. The waste from part IV can be placed in the aqueous waste container.

Questions

1. Explain the reason for the solubility difference observed, between water in petroleum ether, compared to toluene in petroleum ether.

2. Using the Lewis structure of succinic acid, show all of its possible hydrogen bonding interactions with water.

3. From Part II A. explain the reason for the solubility difference observed for cholesterol in water compared to glucose in water. Also, because stearic acid has a carboxylic acid functional group for hydrogen bonding, why doesn't it dissolve in water?

4. From part II B. explain the difference in solubility between citric acid in methylene chloride compared to benzophenone in methylene chloride.

5. From Part IV. write an equation showing why benzoic acid dissolves in NaOH. See equation1. Then write an equation showing what happens when the HCl is added. It is important that you understand this since it is the principle behind "extraction" procedures you will be performing in subsequent experiments.

EXPERIMENT 3

CHROMATOGRAPHY

INTRODUCTION

Chromatography is the most versatile and powerful technique available to a chemist for the purification of organic compounds. It was originally developed by a botanist who used the technique to purify a mixture of various chlorophylls from plant leaves. Because bands of colored pigments were observed during the procedure, it was called chromatography from the Greek word, "chroma," meaning color. Although most chromatographic separations involve colorless mixtures, the name was retained for historical reasons.

Chromatography is the separation of a mixture into its individual components (compounds or ions) via its distribution between two phases. One phase is stationary and the other is moving. The moving ("mobile") phase, that can be a liquid or a gas, carries the mixture through a stationary phase which can be a solid or an immiscible liquid. The components of the mixture interact with the solid phase to varying degrees. Some are more tightly held, some less tightly held by the stationary phase. This partitioning results in separation of the mixture into its individual components that can be visualized as zones along the stationary phase. The result is called a **chromatogram**.

There are a number of different types of chromatography. In **column chromatography** the solid phase is equilibrated with the liquid, mobile phase and held stationary in a vertically mounted, cylindrical metal or glass column. In a process called **sample application**, the mixture is added to the top of the column. The mobile liquid carries the mixture through the stationary phase. The components in the mixture interact with the solid phase to varying degrees and therefore travel through the column at different rates. This process is called **separation**. The pure substances then emerge from the bottom of the column at different times and can be collected in a series of test tubes. The separated solute (dissolved in the mobile solvent) emerging from the bottom of the column is often called the **eluate**. The solvent by itself is called the eluent. In **gas chromatography (GC)** the stationary phase is a solid to which has been adsorbed a layer of nonvolatile liquid. The column is housed in a thermostated oven and the mobile phase is an inert gas such as helium. This mobile phase is called the carrier gas. The sample is applied via a syringe through an injection port to the top of the column. The injection port is heated to a temperature that is high enough to volatilize the sample. Thus the carrier gas moves a gaseous sample through a liquid phase at a high temperature resulting in separation. In **thin layer chromatography (TLC)** the solid phase (often called the **adsorbent**) is applied to the surface of an inert plastic (or glass) sheet. The adsorbent is coated in a thin layer to the plastic support. Using a capillary pipette, a small amount of the mixture to be separated is placed on one end of the plate and dried. This process is called **spotting the chromatogram**. The TLC plate is then placed vertically in a beaker that contains a shallow layer of the liquid mobile phase being used. The mobile phase is often called the **developing solvent**. The sample spot is near the bottom of the plate, but still above the solvent level. The solvent then slowly creeps up the TLC sheet via capillary action. In this process it carries the spotted mixture through the adsorbent causing separation to occur.

Chromatography is also categorized as being preparative or analytical. **Preparative chromatography** involves being able to separate gram quantities of mixtures so that further reactions or analyses can be performed on the separated, purified components. Column chromatography is amenable to preparatory separations. Large columns can be designed to accept a very large amount of the mixture being separated. **Analytical chromatography** involves separating micro- or milligram quantities. The purpose is to simply assess purity of a substance or monitor the progress of a reaction. GC and TLC are common analytical chromatography techniques.

The two common stationary phases in TLC are silica gel and alumina. Both are polar adsorbents that bind polar solutes more tightly than nonpolar ones. Typical interactions are shown below.

Silica Gel interacting with an alcohol   alumina   interacting with an amine

Note that the "very polar" substrates are going to interact more strongly with these adsorbents. This permits us to predict how far various solutes will travel on a TLC plate relative to one another. Table 1 providing the approximate elution sequence of compounds containing various functional groups can be found on the next page. Note that the less polar solutes such as ethers will travel faster than the more polar ones such as alcohols because ethers will not interact as strongly with the alumina or silica gel as the alcohol will. The speed of migration is also affected by the choice of the mobile liquid solvent. A good general rule is that the more polar the solvent, the faster the solutes will migrate up a TLC sheet. One needs to choose a solvent that displays some selectivity in desorbing the solutes from the adsorbent. To achieve adequate separation of the components, a mixture of two or more miscible solvents is often used. This is done to achieve just the right polarity. If the solvent is too polar, all of the components will move

TABLE 1. Migration of Functional Groups on Silica Gel or Alumina

Organic Functional Group	Speed of Migration on TLC Plate
Hydrocarbons	Least adsorbed, will move the fastest
Ethers	
Ketones	
Aldehydes	Intermediate adsorption
Alcohols	
Amines	
Acids, salts	Most highly adsorbed, will move the slowest

at the same speed as the solvent moving up the TLC plate. If the solvent is too nonpolar, the components will remain at the origin where they were spotted. Table 2 lists the elution power of some common TLC solvents used in the laboratory.

TABLE 2. Elution Power of Various Mobile Organic Solvents

Organic Solvent	Ability to Elute Solutes
Petroleum ether / ligroine	Solutes will elute the slowest
Toluene	
Methylene chloride	
Ethyl acetate	
Acetone	
Isopropanol	
Ethanol	
Methanol	
Water	
Acetic acid	Solutes will elute the fastest

Plastic sheets containing a thin layer of silica gel or alumina are commercially available. The sheets are cut into rectangular strips that fit the reservoir beaker containing the mobile liquid solvent. Since the components of many mixtures are colorless, the manufacturer uses an adsorbent impregnated with a dye that fluoresces green when viewed under ultraviolet (UV) light. Wherever there is an organic solute on the sheet, the green fluorescence will be quenched and a brown spot appears. This permits **visualization** of the chromatogram.

In this experiment you will examine two of the most common separation techniques in organic chemistry: column chromatography and thin layer chromatography. Both will be used often in this lab course. The purpose of this experiment is to introduce you to the techniques.

PROCEDURE

Thin Layer Chromatography

1. Your lab instructor will provide you with a rectangular, plastic TLC plate containing a thin layer of silica gel as the adsorbent. You need to handle the sheet as you would a photograph so as not to smudge the silica gel. Use a pencil (not a pen) and a plastic ruler to draw lightly a short line across the width of the plate about 1 cm from the bottom. This line is referred to as the **origin**. Mark off four lanes and label them one through four as shown below. Do not place marks too close to the edge of the plate.

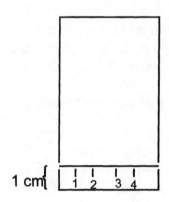

2. Take the TLC to a spotting station. Using the microcapillary tube provided, lightly apply the following four 1% solutions at the origin just above the appropriate mark and number: (1) caffeine, (2) acetaminophen, (3) salicylamide, and (4) a mixture of all three. This is accomplished by holding the microcapillary vertically above the origin then lowering it so that it lightly touches the origin line you drew. Only a small spot of a millimeter in diameter or so is required. Spotting too much solute is called **overloading the chromatogram**. There will be too much solute for the silica gel to handle, equilibration will not take place, and the resulting spot will be too broad after the chromatogram is developed. This phenomenon is called "tailing." Let the four spots dry for several minutes.

3. The mobile organic solvent is ethyl acetate. Take a 250-ml beaker to the hood and add   5 to10 ml of ethyl acetate. Cover the beaker with some aluminum foil and return to your desk. Place the beaker in an area where you can see the contents inside. Once you place the chromatogram in the beaker, it cannot be disturbed. The beaker is now called the **solvent reservoir**.

4. Open the solvent reservoir and place the spotted TLC plate inside. It should be nearly vertical, but not touching the sides of the beaker at the bottom. Again try to handle the TLC sheet as you would a photograph. The bottom of the sheet must be submerged in the solvent; however, the level of liquid at this time must be below the origin as

shown. Gently place the piece of aluminum foil on the beaker. It is important that you not move the solvent reservoir until the experiment is completed.

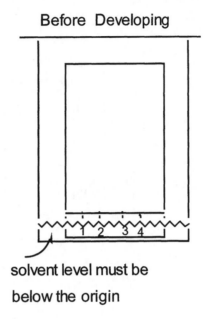

Before Developing

solvent level must be
below the origin

5. From this point on nature takes over. The solvent will slowly ascend the TLC sheet *via* capillary action and develop the chromatogram. You will be able to observe the wetting of the silica gel as this process occurs. The top of this wetting is called the **solvent front**. When the solvent front approaches within 0.5 cm of the top of the TLC plate, open the beaker, remove the sheet, and quickly use a pencil to trace the solvent front before it evaporates. Ethyl acetate is very volatile so you will have only a few seconds to accomplish this. Dry the chromatogram in a hot oven for 5 minutes. You are now ready to visualize and record the spots. Place the chromatogram in the UV box. **NEVER LOOK DIRECTLY INTO THE UV LIGHT AND ALWAYS WEAR SAFETY GOGGLES WHILE USING IT**. Circle all of the brown spots with a pencil. Some organic solutes fluoresce so you may occasionally notice a colored spot. If so, circle and note the color.

6. It is not convenient to save TLC sheets. The adsorbent turns yellow with age and begins to peel away from the plastic backing. Instead, a permanent record of the experiment is accomplished by tabulating reference (R_f) values for all of the components. First, measure the distance in cm. from the origin line to the solvent front line. Then, measure the distance (cm) from the origin to the **center** of each spot. An R_f value for each spot is calculated by dividing this distance by the solvent front distance. An example is diagrammed on the next page. Note that you always obtain a unitless decimal.

7. Repeat the above experiment using 1% acetone solutions of the following three compounds: benzoin, benzil, and hydrobenzoin. The mobile phase in this case will be 70% petroleum ether; 30% ethyl acetate.

8. In the data and results section calculate all of the R_f values and complete Table 3.

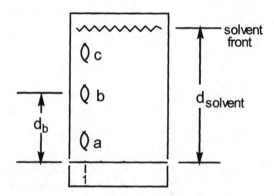

For the imaginary chromatogram above the R_f value for spot b is simply the distance it migrated in centimeters divided by the distance the solvent moved ($R_f = d_b/d_{solvent}$). Separate R_f values would also be calculated for spots a and c.

Microscale Column Chromatography

1. You need to begin by assembling a microscale column (see diagram). Insert the polyethylene frit into the bottom of the column tube. Attach the Leur valve to the frit and secure the apparatus in a clamp. Attach the clamp to a ring stand so that the column is upright and place a beaker under the column. Be sure it is perfectly vertical. Fill the column with ligroin and open the valve to drain part of it into a 50-ml Erlenmeyer flask. This removes air from the fitting and valve. Close the valve after 25% of the ligroin has been drained from the column.

2. The solid phase for this experiment will be alumina. Place the plastic microscale funnel on top of the column and begin to add alumina into the funnel using a microspatula so that the particles slowly pass through the ligroin and settle to the bottom of the column. Continue doing this until the alumina is several centimeters from the top of the surface of solvent. Then open the valve to allow the ligroin to drain into the flask and, at the same time, begin to add more ligroin to the top of the column. At no time should the level of ligroin in the column fall below the level of alumina particles (see diagram). This would permit air bubbles to enter the column. After a few minutes drain the solvent until it is just barely above the solid phase and close the valve.

3. In this experiment you will separate a mixture of two colored dyes, Sudan III (red) and coumarin (yellow), dissolved in methylene chloride. The mobile phase will be 70% ligroin-30% ethyl acetate (EtOAc). Add several milliliters of this solvent to the column and open the valve. Continue to add the ligroin-EtOAc so that the liquid level never drops below that of the solid phase. This process is called "equilibrating the column." After collecting about 40 ml of the mobile phase in the Erlenmeyer flask, drain the column until the mobile phase is just above the top of the solid phase and close the valve. Your alumina column is now equilibrated and ready for "sample application."

4. Using a Pasteur pipet, add about 1 ml of the dye mixture to the column so as not to disturb the top of the solid phase. Open the valve slowly so that the dye moves into the alumina. Then close it. With another Pasteur pipette carefully add a milliliter or so of the ligroine-EtOAc to the column. Open the valve slowly and let it drain until the solvent is just slightly above the top of the alumina. Repeat this two more times. Then completely fill the column with ligroine-ETOAc, open the valve and "develop the chromatogram." You will notice that the mixture will separate into two colored bands as it passes through the column. Continue to add mobile phase to the top of the column.

5. Just before the first colored band begins to exit the column remove the Erlenmyer flask and collect the eluate in a series of test tubes. Allow about 30 drops to enter each test tube before switching to another one. Continue to do this until both colored bands have eluted from the column. Each test tube is called a "fraction" and "separation" is now complete.

6. Record the color of each band and determine their identities.

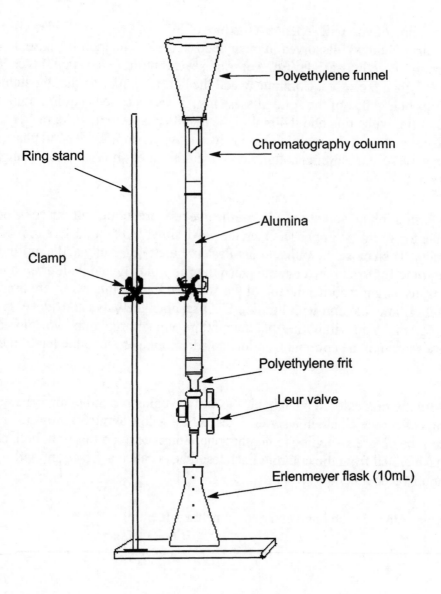

Column chromatography apparatus (microscale)

Microscale Column

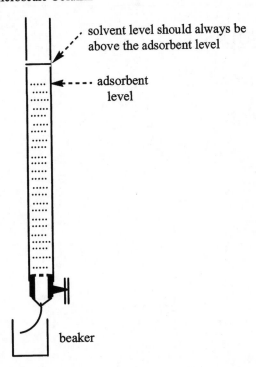

solvent level should always be
above the adsorbent level

adsorbent
level

beaker

Name _____ Sec. _____

DATA and RESULTS

Thin Layer Chromatography

Calculation of R_f values:

Table 3. R_f Values.

solute	R_f	solute	R_f
Acetaminophen		hydrobenzoin	
salicylamide		benzoin	
caffeine		benzil	

Column Chromatography

Table 4. Identity of Solutes Separated by Column Chromatography

Color	Identity of the Band

Name _____ Sec. _____

QUESTIONS

1. Using the Merck Index, draw the structures of benzoin, benzil, and hydrobenzoin. Based on their structures, predict the order of elution in a chromatographic separation on silica gel. Does your experiment verify the prediction? Explain.

2. In performing your TLC experiment, what problem will arise if the level of the developing solvent is higher than the applied spot.

3. Two components have R_f values of 0.2 and 0.6 respectively when TLC is performed on silica gel in 70% CH_2Cl_2 - 30% CH_3OH. Would you expect these R_f values to increase or decrease if the experiment were repeated and the developing liquid was changed to 90% CH_2Cl_2 -10% CH_3OH. Explain.

4. A mixture containing three colored components are added to the top of an alumina column that has been equilibrated with 70% ligroine-30% acetone. As this same solvent is continually added to the top of the column, an orange component elutes in 3 minutes; a blue component in 7 minutes; and a green one in 17 minutes. Which component is presumably the most nonpolar?

   If the entire experiment were repeated with 60% ligroine-40% acetone, would the colored components move slower or faster down the column? Explain.

5. In performing the TLC experiment, what problem will arise if too much of the mixture is spotted on the line near the bottom of the chromatogram?

EXPERIMENT 4

PURIFICATION OF SOLIDS USING MINISCALE AND MICROSCALE CRYSTALLIZATION TECHNIQUES

INTRODUCTION

Earlier in lab you studied the solubility properties of organic compounds in various solvents including water. The organic chemist takes advantage of these properties to purify compounds using a technique called crystallization. A chemical reaction performed in the laboratory normally produces an impure solid (called the **crude product**) that must be purified before further reactions or analyses can be performed. The most common method is to do a **crystallization** in which the crude product is first dissolved in a hot solvent until saturated. Upon cooling slowly, crystals of the desired product form. If there are particles that did not dissolve in the hot solvent, the mixture is filtered while hot, then the filtrate is re-heated and permitted to cool slowly.

The One-Solvent Method

A successful crystallization requires that the desired product be soluble in the hot solvent, but only sparingly soluble in the cold solvent. It is also convenient, but not necessary, if the impurities are a little more soluble than the desired product in the cold solvent. For example, suppose we have 1.5 grams of a crude product that contains 1.2 grams of the desired product, 0.3 grams of an impurity, and the solubility of both in cold ethanol is 0.1 grams/100 ml. Thus the crude product is 80% pure. The mixture is dissolved in 100 ml of hot ethanol. Upon cooling, 0.1 grams of each will remain dissolved producing a product that is 84% pure. Note from the table below that a second crystallization will increase the purification to 91%, and so on.

Table 1

# of crystallizations	Desired Product	Impurity	% pure
	1.2 grams	0.3 grams	80
One	1.1 grams	0.2 grams	84
Two	1.0 grams	0.1 grams	91
Three	0.9 grams	0 grams	100

Although crystallization increases purity, it decreases the yield. In the above imaginary crystallization, the original crude product weighed 1.5 grams. After total purification the overall yield decreased 40% to only 0.9 grams.

Choosing a solvent for performing a crystallization is not always easy. Occasionally one can look up solubility properties in the literature; however, in many cases the trial and error method of experimentation with various solvents is required. Once the correct solvent is found, it is important that the crude product be dissolved in the minimum amount of boiling solvent producing a saturated solution. If undesired particulate matter (or charcoal added for decolorization) is present while doing the mini- or macroscale procedure, the hot solution must be filtered. It is important that the product not crystallize in the filter paper. To prevent crystallization at this point, all of the glassware used must be preheated and placed on a hot steam cone. Gravity filtration is used and the glassware required includes a stemless funnel on an iron ring that is lowered so that it barely touches an Erlenmeyer flask sitting on the steam cone. A specially folded ("fluted") piece of filter paper is used. Your lab instructor will show you how to prepare **fluted filter paper**. It enhances the rate of filtration by providing more surface area through which the solvent can pass. Also fluted filter paper doesn't seat itself tightly against the sides of the funnel. This permits air to enter the Erlenmyer flask causing equalization of pressure between the outside and inside which also increases flow rate.

Two-Solvent Method

In many cases it is not possible to find one solvent with the necessary properties for crystallizing a product. In these instances a two-solvent approach is useful. One finds a solvent in which the solute is soluble and one in which it is only sparing soluble even when hot. Of course, the two solvents must be miscible. As before, a saturated solution of the crude product is prepared. While hot, the second solvent is added dropwise until precipitation just begins. The solution is then slowly cooled to allow crystallization.

Inducing Crystallization

There are instances in which crystallization does not occur upon cooling. It may be that the solution is not saturated. This problem can be overcome by boiling the solution to remove some of the solvent until it is saturated. Occasionally crystallization must be induced and various methods are available. One procedure, called **scratching,** involves taking a glass rod and scraping the inside of the flask. This must be done vigorously in such a way that the rod continually moves in and out of solution. It is thought that "scratching" provides a small amount of solid product on the side of the flask upon which crystallization can begin. Alternately, a genuine crystal of pure product obtained from another source can be added to the solution. This provides a "seed crystal" upon which further crystallization can occur. This process is called **seeding**.

Oiling Out

The purer a desired product is the easier it crystallizes. When the solute is still rather impure, it tends to precipitate as an oily residue. **Oiling out** is also a typical characteristic of low melting point solids. Purification can sometimes be continued by pouring off the supernatant and grinding the oil with a solvent in which it is not soluble. This process is called **trituration** and occasionally cause the oil to solidify. Submitting the triturated solid to the crystallization procedure often results in crystal formation.

Decolorization of the Crystallization Solution

Occasionally a crystallization solution is red, orange, or yellow-colored due to the presence of polar impurities. These can often be removed by adding activated carbon to the hot solution, then filtering though fluted filter paper as discussed earlier. If the filtrate is still slightly colored, the procedure can be repeated. Charcoal has a very high surface area that enables it to adsorb the colored impurities.

Filtration of the Crystals

Once fairly pure crystals are obtained, they are isolated by vacuum filtration on a Buchner funnel (miniscale) or small Hirsch funnel (microscale). The product is then washed with a small amount of ice-cold solvent to remove some soluble impurities still clinging to the crystals. Cold solvent is used because the desired product is somewhat soluble in warm solvent. If the solvent is very volatile, the pulling of air through them by the aspirator is usually enough to dry the crystals so they can be weighed to obtain the yield. Otherwise, the crystals must be scraped into a preweighed watch glass and placed in an oven to dry. It is important that the temperature of the oven not be near the melting point of the desired product.

PROCEDURE

Part 1: Microscale crystallization of organic compounds

In a test tube from your microscale kit, mix about 50 mg of benzil in 1 ml of 95% ethanol. Using a boiling stick to prevent superheating, dissolve the benzil by heating the reaction tube gently over a steam cone. Note the color, and let the solution sit in a beaker at room temperature. Repeat the above procedure in another reaction tube with benzoin. Again, let it sit at room temperature. Note that the benzoin begins to crystallize immediately. Let both reaction tubes sit for about 30 minutes.

Pre-weigh the microscale Hirsch funnel from your kit. Then using this funnel, vacuum filter the benzoin and wash it with a few drops of ice-cold ethanol. Your lab instructor will show you how to set up a vacuum filtration apparatus and connect it to the water aspirator. Be sure all of the components of the apparatus are tightly secured. Remove the funnel from the filtration flask, dry the outside of the funnel with a paper towel, place it in a small beaker, and dry the benzoin in an oven for an hour. Remove the funnel from the oven, allow it to cool, and reweigh. Record the weight and calculate the percent recovery in the data section.

Note that little, if any, of the benzil has crystallized upon standing. As discussed earlier, this means that this solute is too soluble in ethanol. Try using the two-solvent method for crystallizing benzil. Examine the functional groups in benzil and choose a second solvent in which you feel the solute might have limited solubility. Remember this solvent must also be miscible with the first solvent, ethanol. Then re-read the method in the introduction, and crystallize benzil using the two-solvent system. Record your results in the data section. It isn't necessary to filter the crystallized product.

Part 2: Miniscale crystallization of organic compounds

 In this experiment you will crystallize a 1- to 2-gram quantity of an unknown compound containing a red-colored impurity from either water, ethyl alcohol, or a combination of both. You will do some preliminary solubility tests to decide what solvent is adequate and charcoal will be used to decolorize.

Your lab instructor will provide an unknown. Record its letter in the data and results section. Place a milliliter or two of each solvent in separate test tubes and test the solubility of a pea-sized amount of your unknown in each one at room temperature and also while hot. Record your observations below. Then decide what solvent to use in doing the crystallization. Try to use a one-solvent method if possible.

Using a 50-ml Erlenmyer flask dissolve the remainder of your unknown in the hot solvent of choice. Add a microspatula amount of activated charcoal (Norit) and swirl for several minutes while hot. Remove the charcoal from the hot solvent by doing a gravity filtration through fluted filter paper as described in the introduction. Cool the solution to cause precipitation, vacuum filter the product using the macroscale porcelain, Buchner funnel, and wash it with cold solvent. Dry the product and determine a melting point. Identify your product from the list of possible unknowns provided by your lab instructor. Record your results in the data section.

Observations

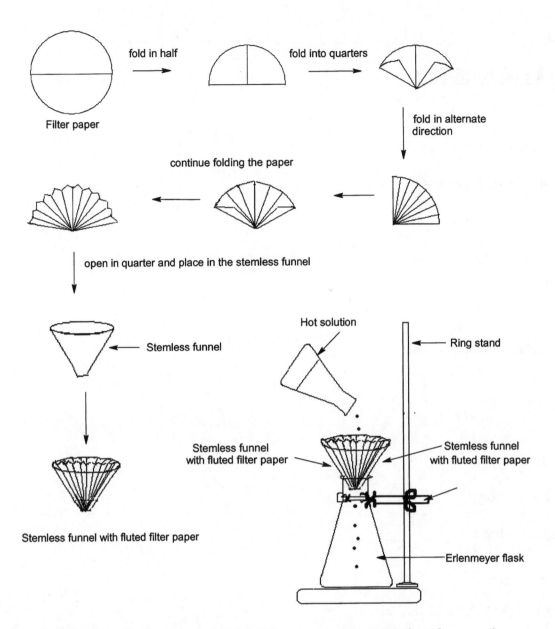

Gravity filtration of hot solution using fluted filter paper and stemless funnel

Name _____ Sec. _____

DATA and RESULTS

Part 1:

Weight of benzoin:

Calculation of percent recovery:

Choice of second solvent:

Part 2:

Letter of unknown _____

Choice of solvent _____

m.p: _____

Identity of unknown _____

QUESTIONS

1. Why do you suppose stemless funnels rather than those with long stems are used when doing a crystallization?

2. During crystallization a person adds too much solvent to the sample and no crystals form upon cooling. How could the desired crystals be obtained at this stage?

3. A student is having problems obtaining crystals from a saturated solution. What are several strategies he can use to cause precipitation to occur?

4. A student needs to crystallize 500 mg of a crude product using the microscale one-solvent method. How does she know how much solvent to use

5. A crude product weighing 2.0 grams contains 0.4 grams of impurities and is therefore 80% pure. The crude product is recrystallized from 100 ml of hot water. Both the desired product and the impurity has a solubility of 0.1 g/100 ml of cold water. What would be the percent purity after two recrystallizations?

EXPERIMENT 5

THE PURIFICATION OF ORGANIC COMPOUNDS USING LIQUID-LIQUID EXTRACTION

INTRODUCTION

Extraction is a method used for separating a desired organic compound from a mixture of various solutes. In liquid-liquid extraction a solution containing various solutes is shaken with another solvent (2) with which it is immiscible. Ideally, the desired solute (p) is extracted into the new solvent leaving the other solutes a,b, and c behind. Solvent 2 is the lower phase because its density is greater than solvent 1.

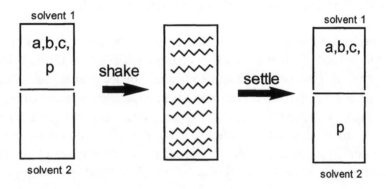

In reality however, an extraction as indicated above is never complete. Although one chooses a solvent 2 in which "p" is more soluble, there will always be some "p" left in the upper phase. Actually "p" will distribute itself between the two phases. From this a **partition coefficient (K)** can be calculated:

$$K = \frac{\text{g/ml of "p" in solvent 2}}{\text{g/ml of "p" in solvent 1}}$$

Let's suppose we have 10.0 grams of a solute dissolved in 300 ml of diethyl ether and we want to extract it into 200 ml of aqueous base where the partition coefficient is 10.0. How many grams of the solute will be extracted? Using the equation above solvent 1 will be the organic phase and it will have 10.0 g /300 ml of ether. Solvent 2 will be the 200-ml aqueous phase with no solute dissolved. After the two phases are shaken and mixed, the solute will partition itself between the two phases. There will then be "x" amount of the solute in the aqueous phase and 10.0 – x amount in the ether.

$$10.0 = \dfrac{\dfrac{x}{200}}{\dfrac{10.0 - x}{300}} \qquad \text{solve for "x"}$$

$$x = 8.7 \text{ grams}$$

Therefore 8.7 grams of the solute has been extracted leaving 1.3 grams (13%) remaining in the ether. To maximize the yield another extraction of the ether layer with 200 ml of aqueous base is performed.

$$10.0 = \dfrac{\dfrac{x}{200}}{\dfrac{1.3 - x}{300}} \qquad \text{solve for "x"}$$

$$x = 1.1 \text{ grams}$$

Another 1.1 grams of solute has been extracted into the aqueous layer leaving 0.2 grams (only 2%) behind. The two 200-ml extracts can then be combined.

As in this example one of the two immiscible solvents used for extractions is usually water. Either the desired product is extracted from an aqueous solution into an organic phase such as ether or it is extracted from an organic phase into an acidic or basic aqueous solution. For this reason it is important to have a knowledge of Bronsted-Lowery acid-base chemistry to understand what is happening in a liquid-liquid extraction procedure. For example, could a saturated solution of sodium bicarbonate be used to extract benzoic acid that was originally dissolved in ether? We need to examine the acid-base neutralization equation and the relative pKa's of the acid and conjugate acid.

$$\underset{\text{pKa} = 4}{C_6H_5COOH} + NaHCO_3 \text{--------}> C_6H_5COO^-Na^+ + \underset{\text{pKa=6.5}}{H_2CO_3}$$

Because the pK_a of the conjugate acid is higher (6.5) than the acid (4), this equilibrium is to the right. Recall from lecture that a Bronsted-Lowery acid-base equilibrium is always on the side of the weaker (more stable) acid. This is just basic thermodynamics. Therefore, the saturated sodium bicarbonate solution will indeed extract benzoic acid from an ether layer. It will end up in the aqueous layer as the sodium benzoate salt. Once the aqueous layer is separated from the organic phase, HCl can be added to the aqueous solution precipitating the benzoic acid which is not very water-soluble. This desired product can then be filtered and dried.

Liquid-liquid extraction can also be used to separate one desired product from another. For example, let's suppose an organic layer such as ether contained benzoic acid and β-napthol (C_9H_7OH) dissolved in it. Will the saturated sodium bicarbonate solution also extract the napthol. To know this we again have to write at the acid-base equilibrium and examine the relative pK_a values.

$$C_9H_7OH \; + \; NaHCO_3 \; \text{--------}> \; C_9H_7O^-Na^+ \; + \; H_2CO_3$$

pKa = 10 (under C_9H_7OH)   pKa=6.5 (under H_2CO_3)

This equilibrium will lie to the left. Therefore, the sodium bicarbonate is not strong enough of a base to extract the napthol. The β-napthol will remain behind in the ether layer as diagrammed below:

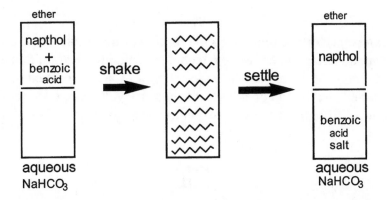

The two layers can be separated and the benzoic acid isolated from the aqueous layer as described earlier. The ether can be poured into a beaker. Since it has a low boiling point of 35°C, it can be evaporated over a steam cone in the hood yielding the β-napthol. Thus, separation of the two desired products can be achieved via liquid-liquid extraction.

In today's lab you will be performing a similar experiment where aqueous base solutions are used to extract organic acids from an organic solvent. In this case the organic solvent is methylene chloride (CH_2Cl_2). To make it easier to observe the extractions, colored acid-base indicators are used. Of course, their colors will vary with pH and solvent. Their structures are given below.

Bromocresol green

a strong acid

I

Alizarin

a weak acid

2

Sudan III

a very weak acid

3

Later in the course you will be doing extractions in which the components are colorless. Using these colored dyes make it easier for you to learn the techniques and observe what is actually happening. The sulfonic acid functional group on the bromocresol green (1) makes it a strong acid. By using aqueous solutions of bases of varying strengths, the bromocresol green and the weak acid alizarin (2) can be selectively extracted from the methylene chloride layer. Mostly Sudan III remains in the methylene chloride.

Extractions can be performed on either a microscale or miniscale level. The equipment and amounts are different; otherwise, the techniques are identical. In this experiment you will learn to do both. Mini- and macroscale extractions are performed using a "separatory funnel" that has an opening at the bottom with a stopcock for draining the lower layer into an Erlenmyer flask or beaker as shown below:

The separatory funnel will fit into a small ring attached to a stand. With the stopcock closed the two immiscible solvents are poured into the funnel through the opening in the top. The plastic cap is placed on the top, the funnel removed from the stand, and with your fingers holding the cap on tightly, it is swirled to enable the two solvents to mix. In many cases a buildup of gas occurs within the funnel and needs to be vented. Venting is accomplished by holding the separatory funnel upside down with fingers firmly on the cap and slowly opening the stopcock. Normally a rush of gas from the stopcock can be heard. The process of shaking and venting is repeated several times. If a carbonate or bicarbonate is present in the extracting solvent, a large quantity of carbon dioxide gas can be produced. Care must be taken not to shake the funnel too much before doing the first few ventings. Usually you will reach a point where no gas forms and the first extraction is completed.

The separatory funnel is then placed in the ring, the top is removed, and the layers are permitted to separate for several minutes. If the extracting solvent containing your product is the lower layer, it is simply drained from the bottom of the funnel by opening the stopcock. **Be sure the top is not on the funnel when you do this**; otherwise, a partial vacuum will build up inside and the drainage will stop. More of this solvent can be added through the top and another extraction performed. This is often repeated a third time. Performing three 100-ml extractions is more efficient than doing only one 300–ml extraction. This will be verified in today's experiment.

If the phase containing your product is the upper one , then simply drain the lower layer from the bottom and pour the remaining upper phase through the top of the funnel into an Erlenmyer flask or beaker. This minimizes contamination of the two layers. The original lower layer can then be placed in the funnel again and the process repeated with more of the extracting solvent.

Microscale extractions require the small test tube (with a cork) or a centrifuge tube with plastic cap found in the kit, as well as a small Pasteur pipette fitted with a latex bulb. For example, the two immiscible phases can be placed in the centrifuge tube. With the plastic cap on, the tube is inverted several times permitting the phases to mix. Venting can be accomplished by slowly opening the cap. Either phase can then be removed using a Pasteur pipette and the extraction repeated several times.

Emulsions

Occasionally the interface between the aqueous and organic layer is not well defined. This is normally due to a colloidal suspension of a small amount of the two phases. This **"emulsion"** can often make it difficult to manually separate the two layers. An emulsion can often be "destroyed" by adding a small amount of a surfactant. This is usually accomplished by adding a few crystals of NaCl or a small amount of a saturated NaCl solution. The water droplets in the colloid will intermolecularly interact with the Na^+ and Cl^- rather than the organic molecules. This helps to break the emulsion. A few drops of a liquid detergent can also be used as a surfactant.

Using Drying Agents

In doing aqueous organic extractions it is also possible that the desired product ends up in the organic layer. It is then isolated by evaporating the organic solvent in the hood. For this to be efficient, there can be no residual water remaining in the organic solvent from the extraction procedure. Drying the organic layer can be accomplished in two ways. It can be extracted with a saturated NaCl solution. Because it is saturated, it helps to draw residual amounts of water from the organic layer. More often a solid **"drying agent"** is added the organic layer and swirled for 10-15 minutes. The **dry** organic layer can then be decanted from the drying agent. Typical drying agents are anhydrous $CaCl_2$ pellets and anhydrous sodium sulfate.

PROCEDURE

The sodium hydroxide solution is caustic and the methylene chloride is a toxic irritant. Avoid skin contact.

Microscale Extraction Experiment

1. Your lab instructor will provide you with a 1×10^{-3} % mixture of three dyes in methylene chloride: bromocresol green, alizarin, and Sudan III. You will find their structures in the introduction. Place two or three milliliters of the mixture in your microscale centrifuge tube. Add 1 ml of 2% aqueous $NaHCO_3$ solution, place the cap on the tube, and mix. The lower layer is the methylene chloride. Allow the phases to separate for several minutes, draw off the upper aqueous layer with a Pasteur pipet, and place it in your 10-ml Erlenmeyer flask. Note the color and record in Table 1. Repeat this extraction three more times. By this time no color should appear in the sodium bicarbonate solution.

2. Using the resulting methylene chloride layer, extract as above with 5% NaOH instead of $NaHCO_3$. Record the colors of the NaOH layer as well as the remaining methylene chloride layer after the second extraction was completed.

3. Based on the relative acidities of the dyes listed in the introduction, predict their identities with respect to the extract and record this in the table also.

Table 1. Microscale Extraction.

	COLOR	**Identity of Dye**
Extract with $NaHCO_3$		
Extract with NaOH		
Remaining CH_2Cl_2 layer		

Extractions Using a Separatory Funnel

1. To practice using a "sep" funnel, place 50 ml of a 1×10^{-4} % solution of alizarin in either a 125-ml or 250-ml funnel. Add 25 ml of the 2% $NaHCO_3$ and extract. Be sure to do the venting correctly and apply the procedure discussed in the introduction to this experiment. Record the color of the two layers and repeat the extraction with another 25-ml portion of "bicarb."

2. Using a fresh 50-ml portion of the alizarin in methylene chloride, extract with one 25-ml aliquot of the 5% NaOH. Again note the color changes and record.

Table 2. Extraction of Alizarin in a Separatory Funnel

	Color of aqueous layer	Color of CH_2Cl_2 Layer
Extract with $NaHCO_3$		
Extract with NaOH		

a. Based on the extraction in the separatory funnel, which one of the two bases is better for extracting alizarin from an organic phase? Explain.

b. Why is the color of the indicator in the aqueous "bicarb" phase different from the color in the aqueous sodium hydroxide?

3. Dissolve two grams of a mixture of *m*-toluic acid and vanillin in about 50 ml of CH_2Cl_2 and transfer to a clean separatory funnel. Extract this aqueous layer with three separate 25-ml portions of 10% sodium bicarbonate solution. Be sure you know what layer is aqueous and use the venting procedure discussed earlier.

4. The product can be isolated from the aqueous $NaHCO_3$ layer in a beaker by adding 6M HCl dropwise with stirring. A precipitate should begin to form. Do this cautiously since the "bicarb" will be converted to CO_2 causing excessive foaming. Continue to add the acid until the mixture is red to blue litmus paper.

5. Place the beaker in an ice bath and stir until cold. Then vacuum filter the product using the macroscale Buchner funnel fitted with filter paper. Wash the product in the funnel with cold water and place it in an oven to dry as you do the work-up on the organic layer.

6. Place the CH_2Cl_2 layer containing the other product in the separatory funnel you have been using, and wash it with 25-ml of saturated NaCl. Which layer is organic and which one is aqueous? Remove the NaCl layer (did you remove the stopper from the top of the funnel?) and pour the organic layer into an Erlenmeyer flask. Add a spatula full of anhydrous sodium sulfate and swirl for about 5 minutes. If all of the sodium sulfate clumps together because of water absorption, add more. Decant the methylene chloride into a 100-ml beaker, add a boiling stick, and evaporate in a hood. Air dry the solid product.

7. Take a melting point of each product and record in the table below.

Table 3. Melting Points of Isolated Products

Product	Melting point (^{o}C)
Product isolated from aqueous layer	
Product isolated from ether layer	

8. Using information from the Merck Index complete Table 4 below by drawing the structures of *m*-toluic acid and vanillin. Also list their literature melting points.

Table 4. Literature Melting Points

Structure	Melting point (^{o}C)

QUESTIONS

1. Comment on the purity of your vanillin and m-toluic acid based on the melting points. If they are not pure, what technique can be used to further purify them?

2. A student adds water to a separatory funnel that already has the product dissolved in an unknown organic phase. After shaking, two layers result. Because the density of the organic phase is unknown, how can the student experimentally determine whether the aqueous layer is the upper of lower one?

3. In doing a microscale extraction in a 5-ml reaction tube, a desired product is dissolved in the bottom layer. Explain experimentally how the layers can be separated.

4. Below are the structures of two organic compounds. Assume they are dissolved in
    ether. You have a saturated aqueous sodium bicarbonate solution as well as 5%
    aqueous NaOH and a separatory funnel. Describe how the two compounds can be
    separated. Write balanced equations showing the reaction occurring during each
    extraction.

5. In a separatory funnel 5.0 grams of a carboxylic acid is dissolved in 200 ml of ether.
    The ether layer is extracted with 100 ml of aqueous $NaHCO_3$ in which K=8.0. How
    many grams of the carboxylic acid will be extracted into the aqueous layer? If the
    carboxylic acid that still remains in the ether layer is extracted with another 100 ml of
    the aqueous bicarbonate, what percent of the carboxylic acid will still remain in the
    ether layer?

EXPERIMENT 6

MICROSCALE FRACTIONAL DISTILLATION

INTRODUCTION

Distillation is a purification procedure in which the components of a mixture of liquids are separated from one another by taking advantage of the differences in their boiling points. For example, if a 50:50 mixture of two liquids are heated to boiling in a "pot," the total vapor pressure above the liquid will equal atmospheric pressure. However, the mixture of the two components in the vapor state will not be 50:50. Instead, the vapor will be richer in the one whose liquid has the lower boiling point. If this vapor mixture is directed away from the pot and caused to condense in the receiver flask, the liquid will now contain a larger quantity of the more volatile liquid. Thus some purification has occurred.

Let's assume we use liquid "A" which has a b.p=110°C, and liquid "B" which has a b.p.=80°C. A 50:50 mixture of the two will boil somewhere between these two temperatures. Let's suppose it boils at 90 degrees. We would like to separate the two liquids so that so the final product contains more of liquid "B" than "A. This can be accomplished by using a simple distillation apparatus. The 50:50 mixture is placed in a round bottomed flask (called the "pot." A **distillation head** containing a **thermometer** is placed on top of the pot in a vertical position. The head will initially collect the vapors and the thermometer will measure the boiling point of the liquid. A **condenser** is attached to the distilling head which directs the vapor away from the pot and toward the receiver. Since the condenser is near room temperature, the vapors will condense and the liquid droplets drip into the **receiver flask**. The liquid droplets comprise the **distillate**.

Let's suppose that the vapor contains 80% of the more volatile liquid "B" as shown in the diagram below. Upon condensation, the liquid will only be 80% pure.

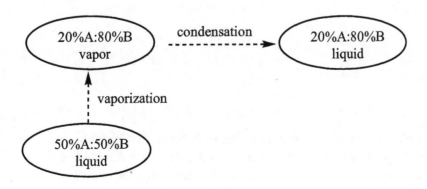

If the difference in boiling points between the two components of the mixture is large enough (usually at least 80°C), then the vapor above the boiling liquid will be mostly that of the more volatile component. Thus most of the lower boiling component will end up in the receiver flask and the higher boiling one will stay in the pot. This is called a **simple distillation**. Normally, this is not the case and some of the higher boiling liquid ends up in the receiver as seen in the previous example. In our example we would need to take the liquid in the receiver flask and subject it to another distillation, followed by another until it was 99% pure.

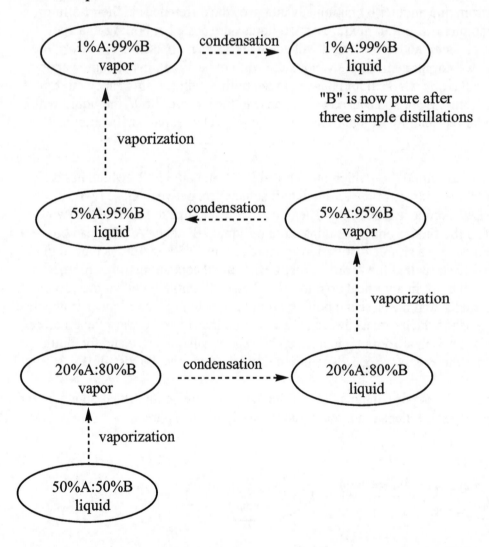

This process would be very time-consuming and inconvenient; therefore, organic chemists utilize what is called a "**fractional distillation**." A fractional distillation is a technique that permits a large number of vaporization-condensation cycles in one operation. This is usually accomplished by packing the vertical distillation head with stainless steel sponge. This turns the head into what is often referred to as a "fractionating column." The sponge provides a large surface area where several successive vaporization-condensations can occur rapidly as the vapor rises. This corresponds to doing a series of re-distillations, and therefore as the vapor rises in the vertical column, it becomes more and more richer in the lower boiling component. The stainless steel sponge is convenient because it is inert, doesn't rust, transfers heat easily, and provides a large surface area without preventing the liquid-vapor mixture from rising. The number of liquid-vapor equilibria that occur in a fractionating column is often referred to as a **theoretical plate**. This refers to the fact that chemical companies using very large columns placed a metal plate (rather than stainless steel sponge) in the column to accomplish the liquid-vapor equilibria. The larger the number of theoretical plates, the more efficient the column would be in separating the components of a mixture.

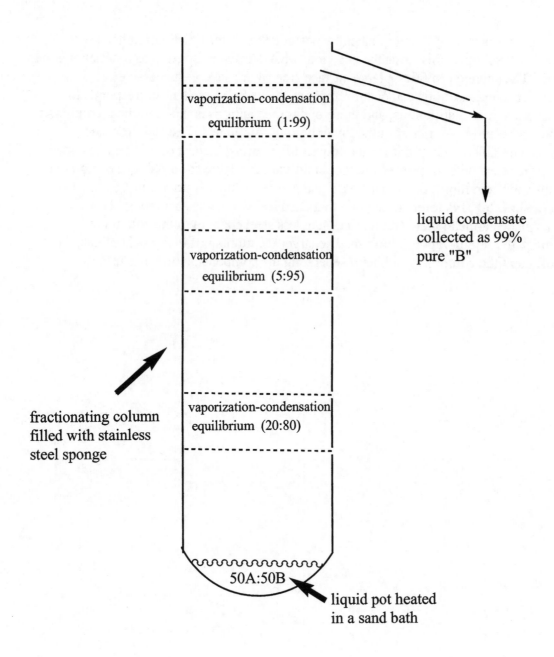

In this experiment you will do a microscale fractional distillation of an unknown containing cyclohexane and n-propyl acetate. There are three possible unknowns containing either 25%, 50%, or 75% cyclohexane. By knowing the boiling points of the two liquids and by plotting temperature vs. volume of distillate collected, you will attempt to decide the percent composition of your unknown.

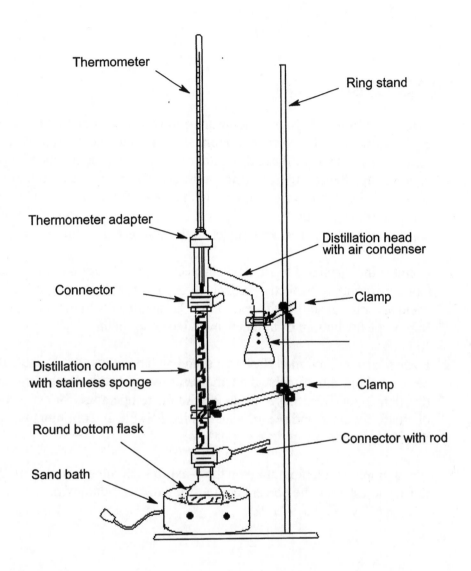

Thermometer

Thermometer adapter

Connector

Distillation column
with stainless sponge

Round bottom flask

Sand bath

Ring stand

Distillation head
with air condenser

Clamp

Clamp

Connector with rod

Microscale Fractional Distillation Apparatus

PROCEDURE

1. Place 4 milliliters of your cyclohexane-propyl acetate unknown and a boiling stone in the short-necked 5 ml roundbottomed flask. Pack the microscale column with stainless steel sponge. Place the metal rod into one of your rubber connectors and use this to attach the condenser to the 5-ml flask. Carefully insert a thermometer into its adapter and attach the thermometer to the top of the microscale Claisen head. Then use your other adapter to connect the Claisen head to the top of the column containing the sponge. Adjust the thermometer so the bulb is below the sidearm of the Claisen.

2. Connect the metal rod to a clamp and secure the entire apparatus on a ring stand so that the column is perfectly vertical. Clamp a 50-ml Erlenmeyer flask to another ring stand and raise it so that the end of the Claisen head is in the neck of the Erlenmeyer. Submerge the bottom of the flask in an ice-water bath.

3. Place the roundbottomed flask in a sand bath and adjust the variac until the mixture begins to boil. Slowly raise the temperature until distillation begins and collect the distillate in the Erlenmeyer flask. Record the temperature for every 4 drops of distillate collected and record your results in Table 1. Continue the distillation until there is less than 0.5 ml in the "pot."

4. Plot a graph with temperature on the y-axis and number of drops on the x-axis. From this try to estimate the percent composition of your unknown.

Name _____ Sec. _____

Data and Results

Table 1. Volume vs. Temperature

Drops	t°C	Drops	t°C

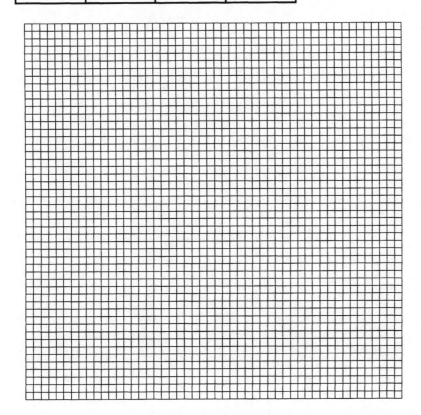

SECTION 2

SYNTHETIC METHODS IN ORGANIC CHEMISTRY

Normally the functional group approach is used in the study of sophomore organic chemistry. The reactions of alkenes, alkyl halides, alcohols, aldehydes, ketones, carboxylic acids, esters, and so on are better understood by seeing how one functional group can be converted to another by using an appropriate synthetic method. In lecture you also learn to use reaction mechanisms to predict regioselective and stereoselective outcomes of organic syntheses. In **Section 2** of the lab manual you will perform experiments to verify that these reactions actually occur as presented in lecture. In doing these experiments you will often use the techniques you learned in **Section 1** to isolate, purify, and identify your products. Also note that a common format is used for performing the experiments and recording the results in this section. It is set up to resemble somewhat the "laboratory notebook" method used by research organic chemists as well as the one you use in general chemistry lab.

You are required to do several things before coming to lab. Before performing any experiment, it is necessary to be aware of the potential hazards involved in using the chemicals and equipment as well as the safety precautions that must be heeded while in the laboratory. This information is usually available in a Merck Index or any of a variety of chemical catalogues from manufacturers of organic compounds. The Aldrich Chemical catalogue is a good choice. Material Safety Data Sheets (MSDS) for compounds are available on the Internet. You need to record this information in the **"Safety/Hazard"** section of each experiment.

You are also required to complete a "**Reactant Table**" before you arrive at lab. A sample can be found in experiment 7. You need to list the names and formula weights of the reactants, reagents, solvent, and products involved in the experiment. For liquids the boiling point (b.p.) and density (D) are listed; for solids the melting point (m.p.) is recorded. This information is usually available from a Merck Index or Aldrich Chemical catalogue. The amount used either as mass or volume is also recorded. From the mass and formula weight, the number of moles can be calculated. This reactant table is simply a convenience. If you take a melting point of your product as a criterion of purity, you only need to check your reactant table to access the literature value. The density of a solvent you use in doing an extraction enables you to determine whether it will be the upper or lower layer. The number of moles of reactants and products is used to calculate the theoretical yield of product.

Before you come to lab, the theoretical yield should be calculated in the "**Data and Results**" section of each experiment. You will begin by writing a balanced equation for the synthesis being performed. Under each structure the grams and moles should be listed with the limiting reagent indicated. Then a theoretical yield is calculated. When the experiment is completed, the percent yield is determined. An example of these calculations can be found below.

As you perform the experiment in the lab, you will record your observations in the appropriate column of the **Procedure/Observation Table**. Observations can include a weight or volume measured; appearance, color, or change in color of a compound, solution or mixture; yield; melting point; and so on. The procedure performed that resulted in the observation is recorded in the "Procedure" column just to the left of the listed observation.

An example of a theoretical yield calculation using the first experiment in this section is shown. The desired product, n-butyl bromide, is synthesized by reacting n-butyl alcohol with sodium bromide and sulfuric acid. The moles of each are calculated by dividing the masses by their respective formula weights. Since the amount of sulfuric acid used is 1.5 ml, its density (1.8 g/ml) is used to find the mass. Thus the limiting reagent is the n-butyl alcohol (0.013 moles).

$$\text{OH} \quad + \text{ NaBr } + \text{ } H_2SO_4 \longrightarrow \text{Br}$$

1.0 g. / 74.1 g/mol = .013 mol   0.015 mol   .028 mol   theoretical yield = 0.013 mol x 137.0 g/mol = 1.8 grams

Using the 0.013 moles, a theoretical yield of 1.8 grams is calculated. This is the maximum amount of product that could be produced in the reaction. In the laboratory the real yield can be substantially less due to loss of material upon using work-up procedures such as an extraction, distillation, chromatographic separation, or crystallization. A typical student yield in this microscale procedure is about 1.0 gram. The percent yield of 56% can then be calculated:

(1.0 grams / 1.8 grams) x 100% = 56% yield

EXPERIMENT 7

THE MICROSCALE PREPARATION OF
n-BUTYL BROMIDE

INTRODUCTION

As you learned in lecture, alkyl halides can be prepared from inexpensive, commercially available alcohols by heating the alcohol with the appropriate hydrogen halide. Typically alcohols do not substitute readily; however, the reaction can be facilitated upon protonation of the alcohol by the hydrogen halide. This converts the alcohol to water which an excellent leaving group due to its low basicity. Because the substrate is a primary alcohol, the reaction proceeds via an **SN2 mechanism**.

Hydrogen bromide is a colorless, corrosive, toxic gas. It is difficult to use safely in an undergraduate lab. A saturated solution of HBr in water containing about 70% HBr by weight will work; however, it is difficult to dispense safely. Instead the HBr is generated directly in the reaction tube by mixing sodium bromide and sulfuric acid. Preparing a reagent his way is referred to as being generated in situ. An excess of H_2SO_4 is used to insure complete conversion of the NaBr to HBr.

Various extraction procedures are used to remove unreacted starting material, reagents, and side products. You need to understand the principles behind these methods by answering the questions at the end of the experiment.

Name _____ Sec. _____ Exp. # _____

Reactant Table

Name of compound	m.p./b.p.	F.W.	Mass (g)	Vol. (ml)	D (g/ml)	moles
Sodium bromide		102.9	1.5			0.015
n-butanol	118°C	84.2	1.0		0.821	0.012
Conc. sulfuric acid		98.08	2.76	1.5	1.84	0.028
n-butyl bromide	104°C	137.03			1.276	

Safety/Hazards

n-butanol is a flammable toxic liquid; avoid skin contact

conc. sulfuric acid is highly corrosive. Upon skin contact, wash contaminated area thoroughly with water.

n-butyl bromide is a flammable liquid, and a toxic irritant. Avoid skin contact.

PROCEDURE

1. Add 1.5 grams of NaBr to the 5-ml, longnecked, roundbottomed flask. Then add 1.5 ml of water and swirl to dissolve most of the solid. Add a boiling stone followed by 1.0 grams of n-butanol. Then carefully and slowly add with swirling, 1.5 ml of concentrated sulfuric acid. Using a connector from your microscale kit, attach the distillation head fitted with a thermometer. Clamp the apparatus to a ring stand and lower the round-bottomed flask into a sand bath. Adjust the heat of the sand so that the mixture slowly refluxes. As directed by your lab instructor, it may be necessary to wrap a wet paper towel around the neck of the flask to minimize evaporation. Reflux the reaction mixture for 30 minutes. Note the layers that form in the reaction flask.

2. Remove the moist towel and begin to distill the product into a small Erlenmyer flask submerged in ice. The major part of the distillate is an azeotrope of water and n-butyl bromide. Continue the distillation until the temperature rises above 120°C. Add about one ml of water to the Erlenmyer flask, mix the contents well, and transfer the mixture to the microscale centrifuge tube. While allowing the layers to separate, dismantle the distillation apparatus, rinse the components with acetone in the hood, and place them in the oven to dry.

3. The n-butyl bromide is the lower layer in your centrifuge tube. Using a Pasteur pipet, remove the lower layer and transfer it to a microscale reaction tube. Add 15 drops of concentrated sulfuric to the reaction tube and mix the layers well by using your finger to flick the tube. The extraction removes side products and unreacted starting material (answer questions #1 and #2 at the end of the experiment). By checking your reactant table, you should be able to predict that the upper layer is the product. To verify this, remove some of the bottom layer with a Pasteur pipet and observe that it is miscible with water.

4. Pasteur pipette carefully remove the top layer and add it to another reaction tube containing two calcium chloride pellets and swirl occasionally as you set up a fractional distillation apparatus. Use only a small amount of stainless steel sponge in the fractionating column. The product sitting in calcium chloride should now be clear. If not, add more and swirl until clear.

5. Using a Pasteur pipette transfer the product (but not the $CaCl_2$) to the distillation pot, add a boiling chip, and purify the n-butyl bromide by doing a final fractional distillation into a pre-weighed receiver flask packed in ice. Use the boiling point you looked up for n-butyl bromide to determine when to begin and end collecting the product.

6. Remove the receiver flask from the ice bath, wipe the outside until thoroughly dry, re-weigh, and record the yield in the data and results section. Also calculate the limiting reagent, theoretical yield, and percent yield.

7. Dispose of waste in the appropriately labeled containers in the hood.

Name _____ Sec. _____ Exp. # _____

PROCEDURE	OBSERVATIONS

Name _____ Sec. _____ Exp. # _____

PROCEDURE	OBSERVATIONS

Name _____ Sec. _____ Exp. # _____

DATA AND RESULTS

Yield of n-butyl bromide(g) _____

Theoretical yield (g) _____

Percent yield _____

Write the equation for this synthesis below and show all of your calculations:

QUESTIONS

1. In this synthesis the crude product was washed with concentrated sulfuric acid. Write an equation showing the reaction that occurs when the unreacted n-butyl alcohol is extracted into the sulfuric acid layer.

2. One of the side products of this synthesis is the production of a small amount of an alkene. Write a mechanism showing how the alkene forms and equation showing why it is extracted into the conc. sulfuric layer when the crude product is washed.

3. What was the purpose of adding the calcium chloride pellets to the crude product before performing the fractional distillation?

4. Write a balanced equation showing what happens when the NaBr is mixed with the H_2SO_4.

5. Why can there be no water present in the n-butyl bromide before the fractional distillation is performed?

EXPERIMENT 8

MICROSCALE PREPARATION OF CYCLOHEXENE

INTRODUCTION

Secondary alcohols can undergo dehydration reactions producing alkenes. This is accomplished by heating the alcohol in the presence of a strong Bronsted acid catalyst like sulfuric or phosphoric acid. The reaction occurs *via* a simple E1 mechanism as described below:

Overall Reaction

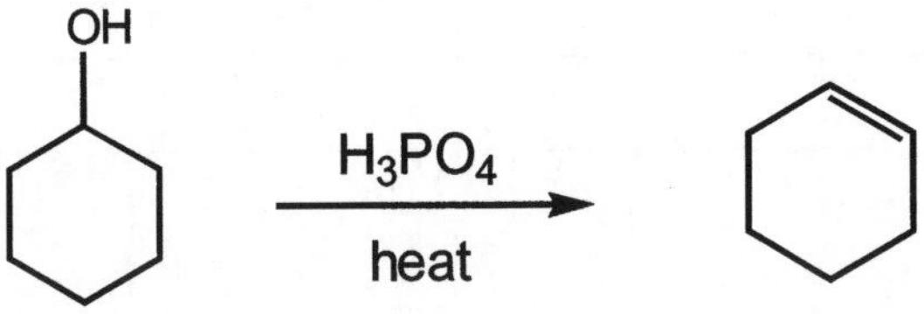

Mechanism

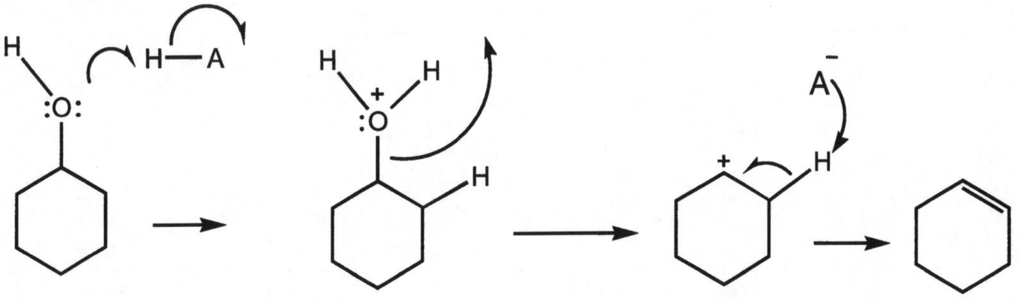

This reaction is highly reversible. The equilibrium is forced to the right by distilling the cycloalkene as it forms. Because water is also in the reaction mixture, the product distills over as an azeotrope. The water is removed and the product purified by doing a fractional distillation. The purity of the product can be assessed by taking an IR spectrum and looking for the absence of the typical IR stretching frequency for alcohols. A stretching frequency for the carbon carbon double bond should also be observed.

Name _____ Sec. _____ Exp. # _____

Reactant Table

Name of compound	m.p./b.p.	F.W.	Mass (g)	Vol. (ml)	D (g/ml)	moles

Safety/Hazards

PROCEDURE

1. Add 3 ml of cyclohexanol to your microscale 5-ml long-necked flask. Then slowly and carefully add 25 drops of 85% phosphoric acid from a Pasteur pipette. Add a boiling chip and swirl for several minutes to mix the reagents. Place some stainless steel sponge in the neck of the flask and attach the distillation head with thermometer.

2. Using a sand bath, distill the product into a microscale Erlenmyer flask sitting in an ice bath. Record the appearance of the distillate and the temperature range for the fractional distillation in the observation section. The temperature should not exceed 100°C. Continue the distillation until there is less than a milliliter of the reaction mixture left in the pot.

3. Using a Pasteur pipette, carefully transfer the distillate to a reaction tube. Add 1 ml of a saturated NaCl solution and perform a microscale extraction. The aqueous NaCl helps to remove water from the organic layer. The fact that it is saturated minimizes the solubility of the cyclohexene in the aqueous layer.

4. Remove the saturated NaCl layer with a Pasteur pipet. It should be the bottom layer; however, after you remove it, be sure you test to see that it is indeed miscible with water. Otherwise it would be the organic layer containing your product. Be sure to record all of your observations.

5. Add 2 or 3 CaCl₂ pellets to the organic layer remaining in the reaction tube and swirl for 5 minutes. If the layer is not clear, add more pellets and repeat the process. As the product is drying, clean and dry your distillation apparatus using acetone, not water.

6. Transfer the clear organic layer to the 5-ml, **short-necked** flask. Add a boiling stone and set up a fractional distillation apparatus with a small amount of stainless steel sponge. Distill the product into a **pre-weighed** receiver submerged in ice. Be sure to record the temperature and appearance of the product.

7. Remove the receiver flask from the ice bath, wipe the outside until thoroughly dry, re-weigh, and record the yield in the data and calculations section. Also calculate the percent yield.

8. Show the product to your lab instructor. If it is clear, you will be instructed to take an IR. Submit the IR, with an interpretation, when you turn in your lab report. Remember that the absence of peaks can be as important as their presence.

9. Dispose of all of the chemicals and waste solutions as indicated by your lab instructor.

Name _____ Sec. _____ Exp. # _____

PROCEDURE	OBSERVATIONS

Name _____ Sec. _____ Exp. # _____

PROCEDURE	OBSERVATIONS

Name _____ Sec. _____ Exp. # _____

DATA AND CALCULATIONS

Yield of cyclohexene(g) _____

Theoretical yield (g) _____

Percent yield _____

Write the equation for this synthesis below and show all of your calculations:

Interpretation of IR:

QUESTIONS

1. What products would be observed upon the dehydration of 1-methyl-1-cyclohexanol? Draw their structures. Predict which one would be the major product.

2. In calculating the percent yield, why isn't it necessary to know the weight or volume of phosphoric acid used?

3. Why can't hydrochloric acid be used to catalyze this dehydration reaction?

EXPERIMENT 9

THE STEREOCHEMISTRY OF THE ADDITION OF BROMINE TO A DOUBLE BOND

INTRODUCTION

As you learned in lecture, the reaction of various reagents with a carbon-carbon double bond can occur via the *syn-* or *anti*-addition mechanism. In this experiment you will study the addition of bromine (Br_2) to the double bond of trans-cinnamic acid and determine whether this is *syn-* or *anti*-addition. The product of the reaction is 2,3-dibromo-3-phenylpropanoic acid. Note that carbons 2 and 3 are chiral. This means that there are four stereoisomers of this compound: the enantiomeric pair (2S,3S), (2R,3R) and the other enantiomeric pair (2S,3R), and (2R,3S). A racemic mixture of one pair results if syn-addition were to occur; a racemate of the other pair results when anti-addition occurs. Because each pair are diastereomers of one another, they would have different melting points. Draw the structures of these compounds in the table below and look up the melting points in the Merck Index. Then answer question #1 at the end of the experiment. In this experiment you will react trans-cinnamic acid with bromine, isolate and purify the product, and determine a melting point. This will enable you to determine whether syn- or anti-addition is the operating mechanism.

2S,3S	2R,3R
Melting point:	
2S,3R	2R,3S
Melting point:	

Name _____ Sec. _____ Exp. # _____

Reactant Table

Name of compound	m.p./b.p.	F.W.	Mass (g)	Vol. (ml)	D (g/ml)	moles

Safety/Hazards

PROCEDURE

1. Place a clean 10-ml Erlenmyer flask containing a stir bar on the analytical balance. Tare the balance and add between 150-200 mg of *trans*-cinnamic acid to the flask and record to the nearest milligram under observations. Add 1.5 ml of methylene chloride, stopper, and swirl the flask to dissolve the cinnamic acid. From the moles of *trans*-cinnamic you used, calculate the number of milliliters of the 1.0 M Br_2 in CH_2Cl_2 needed. You will need to add this amount plus 20% excess.

2. Return to the hood with a 10-ml graduated cylinder and your reaction mixture. Add the calculated amount of the 1.0 M bromine in methylene chloride to the graduate and carefully transfer this to your reaction mixture. **Avoid skin contact with the bromine**. You may want to use disposable gloves. Using a squirt bottle containing alcohol, clean your graduate several times by rinsing into a bromine waste container in the hood. **All of this should be done in the hood.**

3. Place the reaction flask on a stir plate and adjust until the bar slowly rotates. You will soon notice that a chemical change occurred. A precipitate forms, and the color begins to dissipate. Stir for 30 minutes. If the orange-red color disappears, add more Br_2/CH_2Cl_2 and continue stirring.

4. Cool the reaction mixture in an ice bath. Vacuum filter the product into a pre-weighed microscale Hirsch funnel. Wash the product with cold CH_2Cl_2 and dry in the oven for 10 minutes. Determine a yield and take a melting point and percent yield.

5. From the melting point determine whether addition was *syn* or *anti*.

6. Dispose of the product and wash solutions into the appropriate waste containers in the hood.

Name _____ Sec. _____ Exp. # _____

PROCEDURE	OBSERVATIONS

Name _____ Sec. _____ Exp. # _____

PROCEDURE	OBSERVATIONS

Name _____ Sec. _____ Exp. # _____

DATA AND CALCULATIONS

Yield of product (g) _____

Theoretical yield (g) _____

Percent yield _____

Write the equation for this synthesis below and show all of your calculations:

From the evidence you obtained in the experiment, does bromination of a double bond occur via *syn-* or *anti-*addition?

QUESTIONS

1. Using trans-cinnamic acid show mechanistically how both *syn-* and *anti-* addition could potentially occur. Label the R,S configurations of the four, and from the data you collected in your experiment, decide which enantiomeric pair actually formed.

2. What was the purpose of washing the product in the Hirsch funnel with methylene chloride? Why does the methylene chloride have to be cold.

EXPERIMENT 10

MICROSCALE PREPARATION OF CAMPHOR

INTRODUCTION

A typical method of preparing aldehydes, ketones, and carboxylic acids is via the oxidation of alcohols. Oxidation-reduction (redox) reactions occur in concert. One substance undergoes oxidation by donating electrons to another substance that is then reduced. In organic chemistry certain rules are followed in determining whether a particular substrate is being oxidized or reduced:

Oxidation: The loss of hydrogen and/or gain of oxygen by a substrate
Reduction: The gain of hydrogen and/or loss of oxygen by a substrate

Let's look at primary and secondary alcohols as examples.

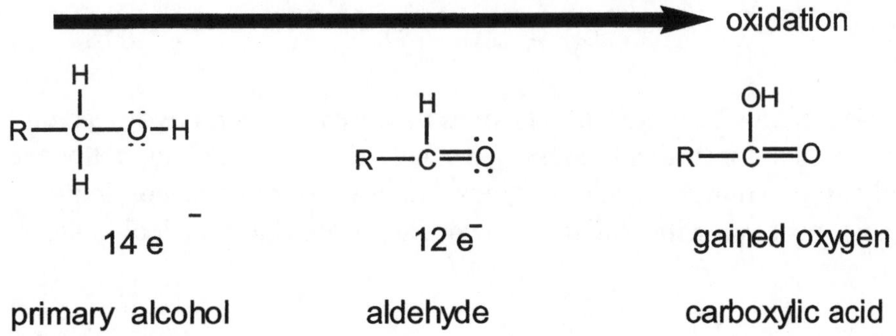

Note that a primary (1°) alcohol can be oxidized to an aldehyde. How do we know this is oxidation? The substrate (1° alcohol) lost two electrons and hydrogen. Note that the conversion of an aldehyde to a carboxylic acid is oxidation because the substrate (aldehyde) lost its remaining hydrogen and gained one oxygen. The substance being reduced in the reactions above are called oxidizing agents. In organic chemistry oxidizing agents are normally inorganic reagents containing transition metals with high oxidation numbers. Typical examples are acidic, aqueous solutions of potassium dichromate ($K_2Cr_2O_7$) where the oxidizing agent is Cr(VI) and alkaline solutions of potassium permanganate ($KMnO_4$) where the oxidizing agent is Mn(VII). Dichromate reagents are orange-colored and upon oxidation of substrates, are reduced to green-colored Cr(III).

Half-Reaction: $Cr_2O_7^{-2} + 14H^+ + 6e^- \longrightarrow 2Cr^{+3} + 7H_2O$
                    Orange   green

This color change can often be used to monitor the oxidation.

In the presence of aqueous potassium dichromate in sulfuric acid primary alcohols are completely oxidized to the carboxylic acid. The second oxygen in the acid product originates from the water.

$$R-\overset{\overset{\displaystyle H}{|}}{\underset{\underset{\displaystyle H}{|}}{C}}-O{\cdot}H \xrightarrow[\text{H}_2\text{O}]{\text{K}_2\text{Cr}_2\text{O}_7 \,/\, \text{H}_2\text{SO}_4} R-\overset{\overset{\displaystyle OH}{|}}{C}=O$$

To stop the oxidation of primary alcohols at the aldehyde stage, a nonaqueous Cr(VI) reagent is needed. A typical reagent of this type is pyridinium chlorochromate (PCC) dissolved in a nonpolar, aprotic solvent such as methylene chloride (CH_2Cl_2).

PCC

$$R-\overset{\overset{\displaystyle H}{|}}{\underset{\underset{\displaystyle H}{|}}{C}}-O{\cdot}H \xrightarrow[\text{CH}_2\text{Cl}_2]{\textbf{PCC}} R-\overset{\overset{\displaystyle H}{|}}{C}=O$$

Secondary alcohols can only be oxidized to ketones because there is no hydrogen on the carbonyl carbon to be lost for further oxidation to occur. Even so, PCC in methylene chloride is usually used to minimize side reactions that can occur in aqueous acidic media. In this experiment you will oxidize the secondary alcohol, isoborneol to the ketone, camphor.

ISOBORNEOL   CAMPHOR

Natural camphor is the dextrorotatory isomer obtained from a tree indigenous to China and Japan. Most camphor is synthesized in the laboratory and consists of the +/- racemate. Noted for its penetrating odor, it is used therapeutically as a topical anti-infective and as a moth repellant.

Name _____ Sec. _____ Exp. # _____

Reactant Table

Name of compound	m.p./b.p.	F.W.	Mass (g)	Vol. (ml)	D (g/ml)	moles

Safety/Hazards

PROCEDURE

The reaction mixture is set up and allowed to sit until the next lab period.

1. In a 10-ml Erlenmyer flask from your microscale kit dissolve about 150 mg of isoborneol in 3 ml of methylene chloride (CH_2Cl_2). Add 300 mg of PCC and swirl for about 10 minutes. Note any color changes. Try to minimize the clumping of the PCC on the bottom of the flask. Place a cork (not a rubber stopper) on the Erlenmyer and secure it tightly with a piece of Parafilm to prevent evaporation of the CH_2Cl_2. Place the flask in the tray provided by your lab instructor.

2. At the beginning of the next lab period remove the Parafilm/cork and use a stirring rod to break apart the clumps on the bottom of the flask. Vacuum-filter the mixture through the microscale Hirsch funnel.

3. Pour a microscale chromatography column ¾ full with a slurry of alumina in CH_2Cl_2. Equilibrate the column with CH_2Cl_2, then add the filtrate from your vacuum filtration to the top of the column. Elute the column with CH_2Cl_2 and collect 40 ml of the eluate in a 50-ml pre-weighed beaker. This will separate the camphor from dissolved chromium salts as well as any unreacted isoborneol.

4. Evaporate most of the methylene chloride on a steam cone in the hood. Be sure the steam is low enough that water vapor does not condense on the inside of the beaker. The final milliliter or two of CH_2Cl_2 can be removed with a stream of nitrogen gas to produce the solid product. Carefully wipe the outside of the beaker and re-weigh to obtain the yield of the product. Take an IR of the product.

5. Calculate a percent yield.

6. Dispose of the methylene chloride in the halogenated waste container. PCC should be placed in the chromium waste container.

Name _____ Sec. _____ Exp. # _____

PROCEDURE	OBSERVATIONS

Name _____ Sec. _____ Exp. # _____

PROCEDURE	OBSERVATIONS

Name _____ Sec. _____ Exp. _____

DATA AND CALCULATIONS

Yield of camphor:

Percent yield:

Write the equation for this synthesis below and show all of your calculations.

Write a short discussion comparing the presence or absence of alcohol and carbonyl stretching frequencies in the IR with respect to the purity of the product.

Name _____ Sec _____ Exp. _____

QUESTIONS

1. Apparently isoborneol binds more strongly to the alumina than camphor. Explain this behavior based on the structure of isoborneol.

2. As mentioned in the introduction, PCC is the preferred reagent for the oxidation of secondary alcohols because sulfuric acid and water can cause side reactions. What intermediate do you suppose is involved in these side reactions? Show how this intermediate forms using cyclohexanol, H_2SO_4, and water.

3. Indicate whether the reactions below are oxidations or reductions. Explain your decision in each case.

KMnO₄ ⟶ MnO₂

EXPERIMENT 11

THE MICROSCALE REDUCTION OF BENZIL
WITH SODIUM BOROHYDRIDE

INTRODUCTION

Oxidation-reduction (redox) reactions occur in concert. One substance undergoes oxidation by donating electrons to another substance that is then reduced. In organic chemistry certain rules are followed in determining whether a particular substrate is being oxidized or reduced.

Oxidation: The loss of hydrogen and/or gain of oxygen by a substrate
Reduction: The gain of hydrogen and/or loss of oxygen by a substrate

When an organic substrate is reduced, the reagent being oxidized is called the **reducing agent.** A common reagent for reducing aldehydes and ketones to alcohols is sodium borohydride. Normally this is carried out in a protic solvent such as water, methanol, or ethanol. An equivalent of hydride (H:) is provided to the substrate as visualized below:

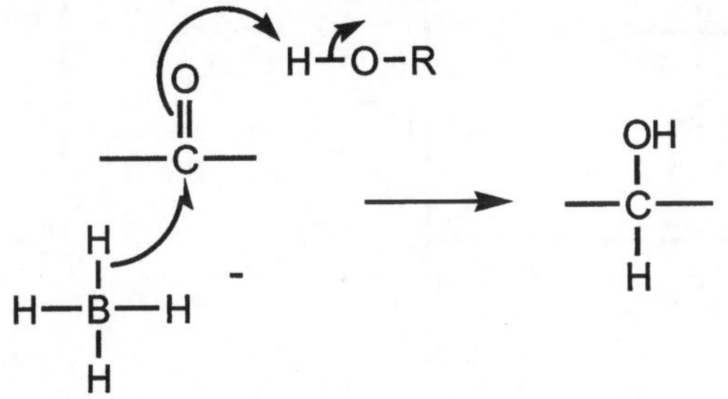

In this experiment you will study the reduction of a diketone, benzil, by $NaBH_4$. Before beginning the experiment you need to look up the structure and melting point of benzil as well as hydrobenzoin, the product. Because hydrobenzoin has two chiral carbons, various stereoisomers exist, and the Fischer projections of all of these compounds should be drawn in the table on the next page. The identity of the correct diastereomer of hydrobenzoin that results can be determined from its melting point and verified by its R_f value on a silica gel TLC plate.

Name _____ Sec. _____ Exp. _____

Benzil

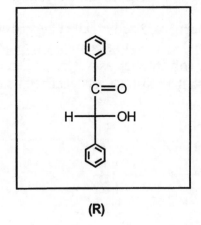

Benzoin

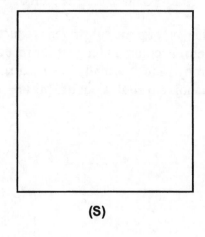

(R)

(S)

meso-**hydrobenzoin**

(1R,2R)

(1S,2S)

racemic   hydrobenzoin

Name _____ Sec. _____ Exp. # _____

Reactant Table

Name of compound	m.p./b.p.	F.W.	Mass (g)	Vol. (ml)	D (g/ml)	moles

Safety/Hazards

PROCEDURE

1. Dissolve 75 mg of benzil in 20 drops of 95% ethanol by stirring in a reaction tube. Place the tube in an ice bath and continue to stir for several minutes.

2. Then add 15 mg of sodium borohydride and stir. Note the changes in color as you perform this procedure. After 5 minutes, remove the tube from the ice bath and continue to stir for another 5 minutes. At this time the mixture should be nearly colorless.

3. Add about 10 drops of water to the reaction mixture and place it in a sand bath next to another test tube containing just water. When hot, add some of the boiling water to the reaction mixture until it just begins to turn cloudy. This suggests that the alcohol-solvent mixture is now saturated.

4. Let the mixture cool to room temperature slowly. Do not stir.

5. Place the reaction tube in an ice bath and stir for 5 minutes. Then filter the product into a microscale Hirsch funnel. Wash the hydrobenzoin with small amounts of ice-cold 50% ethanol-water. Remove a small amount to be dissolved in acetone for TLC. Place the remainder of the product in the oven to dry.

6. Spot some of your product in the middle lane of a fluorescent silica gel sheet. In four other lanes spot authentic samples of benzoin, benzil, *meso*-hydrobenzoin, and *racemic*-hydrobenzoin provided by your lab instructor. Develop the chromatogram in 70% petroleum ether-30% ethyl acetate. Calculate and record the R_f values in the data and calculations section.

7. Remove the product from the oven, cool to room temperature, and weigh. Record the yield and calculate a percent yield. Show the product to your lab instructor, and determine a melting point.

8. Dispose of all washings in the appropriate waste bottle in the hood.

Name _____ Sec. _____ Exp. # _____

PROCEDURE	OBSERVATIONS

Name _____ Sec. _____ Exp. # _____

PROCEDURE	OBSERVATIONS

DATA AND CALCULATIONS

1. Diagram your chromatogram and indicate the location of the components. Determine the R_f values for the spots.

2.

   Yield of product:

   m.p. of product:

   percent yield:

3. Based on the m.p. and R_f value, decide whether the product is *meso*-hydrobenzoin or racemic-hydrobenzoin. Defend your choice.

QUESTIONS

1. Why does benzil have a higher R_f value than hydrobenzoin?

2. Why don't the (2S,3S)- and (2R,3R)-hydrobenzoins separate on TLC?

3. You probably observed that the R_f values for the meso and racemic isomers were nearly identical. However, careful TLC has shown that they do have slightly different values. How are these two related, and why do they have different R_f values.

4. A large excess of $NaBH_4$ was used to force the reduction of benzil to the right. Even so, you may have observed only modest yields. Propose a reason for this based on the subsequent work-up you used.

EXPERIMENT 12

THE MICROSCALE SODIUM BOROHYDRIDE REDUCTION OF VANILLIN

INTRODUCTION

Oxidation-reduction (redox) reactions occur in concert. One substance undergoes oxidation by donating electrons to another substance that is then reduced. In organic chemistry certain rules are followed in determining whether a particular substrate is being oxidized or reduced.

Oxidation: The loss of hydrogen and/or gain of oxygen by a substrate
Reduction: The gain of hydrogen and/or loss of oxygen by a substrate

When an organic substrate is reduced, the reagent being oxidized is called the **reducing agent.** A common reagent for reducing aldehydes and ketones to alcohols is sodium borohydride. Normally this is carried out in a protic solvent such as water, methanol, or ethanol. An equivalent of hydride (H:) is provided to the substrate as visualized in the introduction to experiment.

The substrate in this reaction is vanillin (4-hydroxy-3-methoxybenzaldehyde). This is the naturally occurring component found in "vanilla" that provides its characteristic taste and odor. The aldehyde functional group can be reduced to a primary alcohol as shown below:

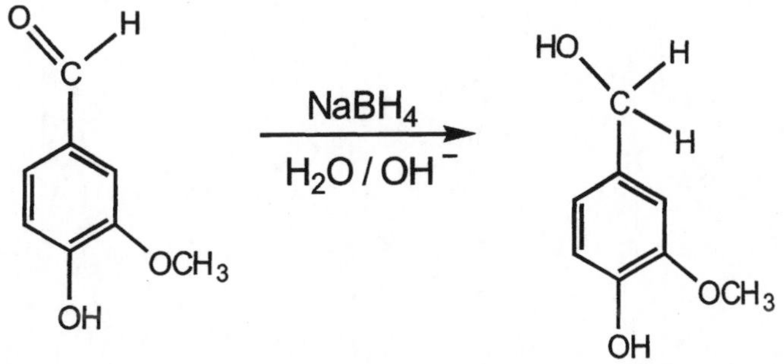

The product is vanillyl alcohol (4-hydroxy-3-methoxybenzyl alcohol). Note that the substrate has accepted a hydride (H:) from $NaBH_4$ and a proton from the solvent. The pK_a of a hydroxyl group attached to a benzene ring (often called a "phenolic –OH") is much lower than the aliphatic alcohol functional group. This enables vanillin to dissolve in sodium hydroxide permitting the reduction to occur.

Name _____ Sec. _____ Exp. # _____

Reactant Table

Name of compound	m.p./b.p.	F.W.	Mass (g)	Vol. (ml)	D (g/ml)	moles

Safety/Hazards

PROCEDURE

1. Weigh 300 mg of vanillin into a 10-ml Erlenmyer flask equipped with a stirring bar. Add 4 ml of a 0.5M NaOH solution and stir until all of the starting material dissolves. If it doesn't completely dissolve, add another drop of the sodium hydroxide solution. Note the color of the solution and place the flask in an ice bath.

2. Add 50 mg of sodium borohydride to the reaction mixture in approximately 10 mg portions over a period of about 5 minutes. Bring the mixture to room temperature and let it stir until almost clear. This usually takes about 20 to 30 minutes.

3. Place the flask in the ice bath and neutralize the NaOH by slowly and cautiously adding a 3M HCl solution drop-wise. Note any observations. Swirl the flask well after each addition. After about 1 ml of the HCl is added, check with litmus paper to see if it is acidic. Continue to add 3M HCl until the reaction mixture is acidic to blue litmus paper.

4. While the mixture is still in the ice-water bath, scratch the side of the Erlenmeyer flask to induce crystallization of the product. Continue to scratch and swirl until precipitation is complete. This usually takes about 10 minutes.

5. Vacuum filter the product into a Hirsch funnel and wash it with five 2-ml portions of ice-cold water. Place the funnel in a 90°C drying oven for 20 minutes.

6. Remove a small amount and take a melting point, but leave the funnel in the oven for another 20 minutes.

7. Weigh the product and record the yield.

8. Take an IR of the product.

9. If the IR and melting point indicate that the product is not pure, recrystallize it from ethyl acetate.

Name _____ Sec. _____ Exp. # _____

PROCEDURE	OBSERVATIONS

Name _____ Sec. _____ Exp. # _____

PROCEDURE	OBSERVATIONS

Name ____ Sec. _____ Exp. _____

DATA AND CALCULATIONS

Yield of vanillyl alcohol:

Calculate your percent yield. Hint: Use your textbook to find the stoichiometry of this reaction before determining the limiting reagent.

Write a short discussion comparing the presence or absence of alcohol and carbonyl stretching frequencies in the IR with respect to the purity of the product.

Name _____ Sec. _____ Exp. _____

QUESTIONS

1. Draw the structure of vanillin and write a balanced equation showing its reaction with sodium hydroxide.

2. After the reduction is complete, HCl is added to precipitate the crude product. Write a balanced equation showing this reaction.

3. What impurities are removed from the product upon washing it with ice-cold water?

EXPERIMENT 13

THE MINISCALE SYNTHESIS OF BENZOCAINE BY DIRECT ESTERIFICATION

INTRODUCTION

In this experiment you will study the conversion of the carboxylic acid functional group to one of its derivatives, the ester. There are a number of ways of doing this conversion. A more indirect approach involves converting the carboxylic acid to the more reactive acid chloride. This is accomplished with the reagent thionyl chloride as shown below. The chloride in this functional group is easily replaced at room temperature with a variety of nucleophiles including an alcohol producing the ester.

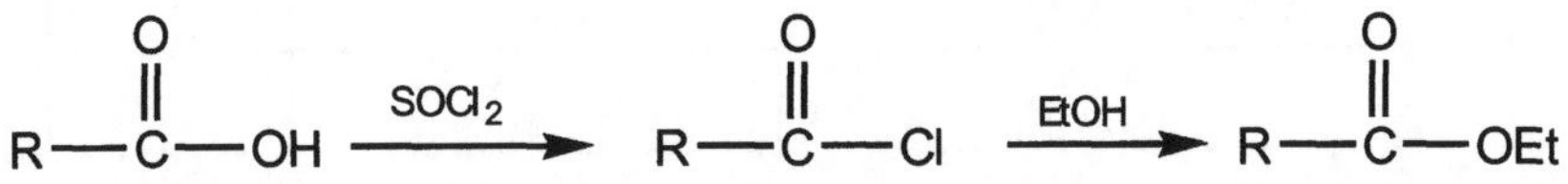

In this experiment the local anesthetic, benzocaine, will be synthesized from p-aminobenzoic acid via direct esterification with ethanol. Because the hydroxyl is not a very good leaving group, the reaction needs to be acid-catalyzed at a higher temperature. Benzocaine is not water soluble, and because it does not suffuse well into tissue, is often used in skin ointments. It is the ingredient found in many sunburn preparations.

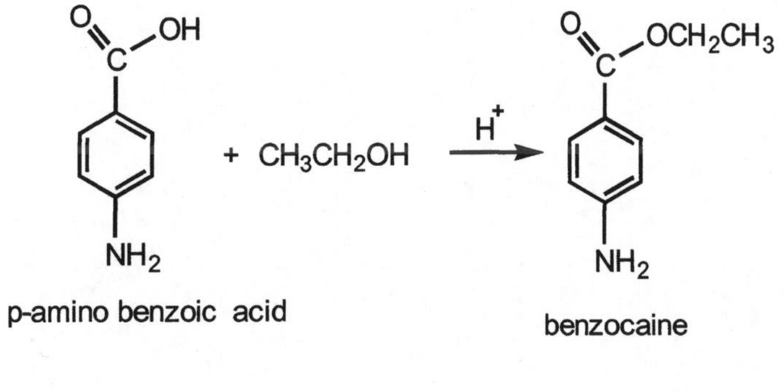

p-amino benzoic acid   benzocaine

Name _____ Sec. _____ Exp. # _____

Reactant Table

Name of compound	m.p./b.p.	F.W.	Mass (g)	Vol. (ml)	D (g/ml)	moles

Safety/Hazards

PROCEDURE

1. Mix 1.5 grams of p-aminobenzoic acid with 35 ml of absolute ethanol in a 100-ml round-bottomed flask. Swirl the mixture until most of the starting material dissolves. Take the flask to the hood and carefully add 40 drops of the concentrated sulfuric acid and swirl. Record your observations.

2. Add several boiling chips and begin to set up a reflux apparatus with a water-jacketed condenser. Be sure to connect the water inlet hose from the faucet to the bottom of the condenser. Place the flask in a heating mantle, attach the condenser to the top of the flask using a small amount of vacuum grease to seat the fitting, and secure the apparatus to a ring stand as directed by your lab instructor. Adjust the variac attached to the heating mantle until the reaction mixture begins to boil.

3. Within 10 minutes, all of the solid should dissolve. If not, add more acid a drop at a time until it does. At the option and direction of your instructor, you may have to add more ethanol. Unless all of the solid dissolves, benzocaine will not form. Reflux for a total of 60 minutes. Your lab instructor will show you how to occasionally swirl the reaction mixture by rotating the entire ringstand.

4. Let the reaction mixture cool for about 5 minutes, remove the reflux condenser, and place the flask on a cork ring. Transfer the reaction mixture to a beaker and add about 40 ml of crushed ice to the reaction mixture.

5. To neutralize the mixture add a 2M Na_2CO_3 solution drop-wise with stirring. Record what happens when you add the base. Add the Na_2CO_3 to the mixture until red litmus paper turns blue. It shouldn't require more than 20 ml of the sodium carbonate solution. If foaming occurs, then you haven't added enough.

6. If precipitation of the product has not occurred, some crushed ice should be added to the mixture. Scratching the side of the beaker with a stirring rod may help to induce crystallization. Cool the reaction mixture in an ice-water bath to complete precipitation of the product.

7. Collect the product in a Buchner funnel using vacuum filtration. Wash it thoroughly with three 15-ml portions of water. Let the product dry in your lab drawer until next week.

8. Weigh the product and determine a melting point. If the melting point indicates that the benzocaine is fairly pure, show the product to your lab instructor. Otherwise, recrystallize using the two-solvent method you devised when answering question 1.

Name _____ Sec. _____ Exp. # _____

PROCEDURE	OBSERVATIONS

Name _____   Sec. _____   Exp. # _____

PROCEDURE	OBSERVATIONS

DATA AND CALCULATIONS

Yield of benzocaine:

m.p. :

Calculation of percent yield:

Discuss your assessment of the purity of your product based on its melting point.

Name _____ Sec. _____ Exp. # _____

QUESTIONS

1. The benzocaine could be purified by recrystallization. Based on its solubility in this experiment, propose a two-solvent method of recrystallizing about one gram of benzocaine. Then describe how you would carry out the recrystallization. Specify temperatures, techniques, and equipment used.

2. Write the equation that shows what happens when precipitation occurs as the sulfuric acid is added to the p-aminobenzoic acid in ethanol at the very beginning of the experiment.

EXPERIMENT 14

THE MINISCALE SYNTHESIS AND PURIFICATION OF ASPIRIN

INTRODUCTION

Many over-the-counter analgesic and antipyretic drugs contain aspirin. Some include additional ingredients such as buffers to reduce acidity, stimulants such as caffeine, and other analgesics such as acetaminophen. In today's experiment you will synthesize aspirin from salicylic acid. Salicylic acid is a bifunctional aromatic compound containing the alcohol and carboxylic acid functional groups. It reacts with acetic anhydride in the presence of an acid catalyst forming acetylsalicylic acid, commonly called "aspirin."

Because of its bifunctionality, various polymeric side products can also form as the reaction proceeds. These are typically water-insoluble and can be removed during the aqueous sodium bicarbonate treatment in the work-up of the crude product.

TLC or melting point could be used as a criterion of purity. Another method available is a colorimetric assay. The most likely impurity in the final product is the unreacted salicylic acid. Salicylic acid, like most phenols, forms a highly colored complex with Fe^{+3} ion. Aspirin does not have a phenol group and should not react. Thus, by treating your product with a ferric chloride solution, the presence or absence of the phenol group can be detected and used to assess the purity of the aspirin.

Name _____ Sec. _____ Exp. # _____

Reactant Table

Name of compound	m.p./b.p.	F.W.	Mass (g)	Vol. (ml)	D (g/ml)	moles

Safety/Hazards

PROCEDURE

1. Rinse a 125-ml Erlenmeyer flask and 10 ml graduated cylinder with acetone and place them in the drying oven before prelab lecture.

2. Weigh 2.0 grams of salicylic acid and transfer the solid to the Erlenmeyer flask. Add 5.0 ml of acetic anhydride and swirl. Then add 5 drops of concentrated sulfuric acid from the plastic dropper bottle and swirl the flask. The salicylic acid should dissolve. Heat the flask on a steam cone for about 15 minutes. Allow the flask to cool to room temperature. The aspirin should crystallize into a solid mass. If it does not, scratch the walls of the Erlenmeyer in an ice bath to induce crystallization. If crystals don't form it is likely that either the glassware or chemicals were wet. Begin the reaction again with dry equipment.

3. After crystallization is complete, add about 50 ml of water, cool in an ice bath, and stir well. Collect the product *via* vacuum filtration into your large, Buchner funnel. Be sure the filtration apparatus is clamped tightly. Use cold water to transfer all of the aspirin from the Erlenmeyer to the funnel. Then wash the product in the funnel with small portions of cold water. The water will convert most of the excess acetic anhydride to acetic acid. The cold water will dissolve the acetic acid which ends up in the filtration flask that holds the Buchner. Set a small amount of the crude aspirin aside for later tests and recrystallize the remainder according to the following procedure.

4. Transfer the crude product to a 150-ml beaker, add 25 ml of saturated sodium bicarbonate solution, and stir. The aspirin should dissolve. Some of the solid may not dissolve. This is a polymeric product that occasionally forms during the esterification process. Filter through a Buchner funnel, wash the beaker and funnel with 10 ml of water. **Remember your product is in the filtration flask, not the funnel**. Prepare a mixture of 3.5 ml of conc. HCl and 10 ml of water in a 150-ml beaker. Transfer the filtrate slowly with stirring to the 150-ml beaker. Be careful because foaming may occur due to CO_2 formation. The aspirin should precipitate. Test the mixture with blue litmus paper. If it is not acidic, add conc. HCl dropwise **with stirring** until it is acidic. If your product does not precipitate, scratch the sides of the Erlenmeyer as it sits in the ice bath. If all else fails, add a seed crystal of pure aspirin. This will normally induce crystallization.

5. Cool the mixture in an ice bath, vacuum filter through a Buchner, and wash well with ice-cold water to remove the excess HCl. It is essential that the water be ice-cold to minimize hydrolysis of the aspirin. Scrape the product into a watchglass and store in your drawer until next week. Weigh the product and determine a percent yield.

6. Add 5 ml of water to four separate test tubes. Add a few crystals of phenol to tube 1, salicylic acid to tube 2, a few crystals of the crude product you set aside to tube 3, and a few crystals of the pure product to tube 4. Then add about 10 drops of 1% ferric chloride solution to each one. Note and record the colors that form under Observations. Formation of an iron(III)-phenol complex will give various red to violet colors depending on the particular phenol present.

Name _____ Sec. _____ Exp. # _____

PROCEDURE	OBSERVATIONS

Name _____ Sec. _____ Exp. # _____

PROCEDURE	OBSERVATIONS

DATA AND RESULTS

Weight of product:

Calculation of percent yield:

Discuss the results of the ferric chloride test with respect to the purity of your product:

QUESTIONS

1. Why must the Erlenmeyer flask and graduated cylinder be dried in an oven before the reaction is run?

2. Write an equation showing why the crude aspirin dissolves in the saturated $NaHCO_3$ solution.

3. Write an equation showing why the aspirin precipitates from when HCl is added to the saturated $NaHCO_3$ solution.

EXPERIMENT 15

THE MINISACLE HYDROLYSIS OF METHYL SALICYLATE

INTRODUCTION

Methyl salicylate is a natural product that was originally isolated 150 years ago via extraction of a wintergreen plant. For this reason it is often called "oil of wintergreen." It exhibits the typical wintergreen odor and taste and is often used as a flavoring agent. It is also used as a rubbing liniment because it exhibits analgesic properties and is rapidly absorbed through the skin. As you can see from its structure below, methyl salicylate contains the "ester" functional group. In this experiment you will verify that esters can be hydrolyzed to carboxylic acids and alcohols under alkaline conditions. This reaction is often referred to as **saponification** ("soap making") because it was originally applied to the alkaline hydrolysis of fats to make soap.

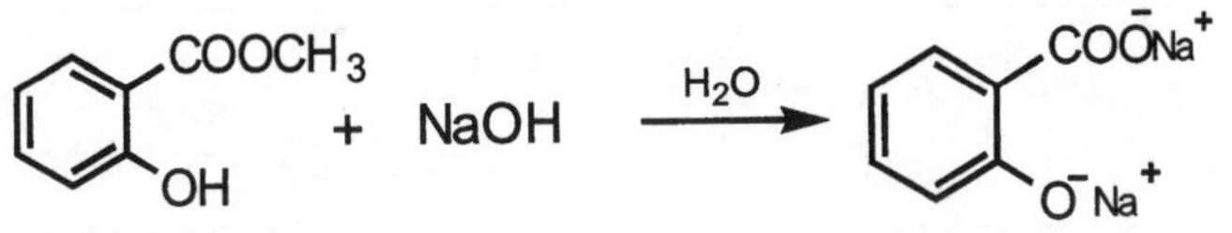

methyl salicylate   disodium salicylate

Because the reaction mixture is in a strong base, the product that forms is the disodium salt of salicylic acid, and is, of course, water soluble. This salt can be protonated by adding a strong acid to neutralize the excess NaOH. This precipitates the "free" salicylic acid that is very insoluble in cold water.

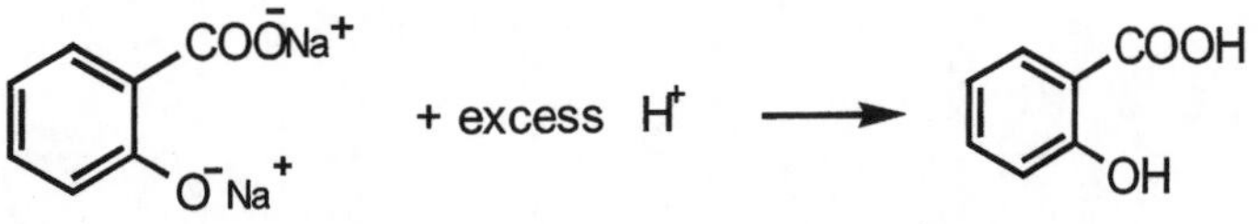

disodium salicylate   salicylic acid

The product can then be filtered and rinsed free of excess acid and inorganic salts with cold water. However, the water rinsing will not remove any unreacted methyl salicylate. Therefore, the product is recrystallized from hot water.

Name _____ Sec. _____ Exp. # _____

Reactant Table

Name of compound	m.p./b.p.	F.W.	Mass (g)	Vol. (ml)	D (g/ml)	moles

Safety/Hazards

PROCEDURE

1. Dissolve 5.0 grams of NaOH in 25 ml of water in a 50-ml roundbottomed flask. Be careful because NaOH is very caustic and the reaction is highly exothermic. Make sure there are no cracks in the flask. Cool the solution in an ice bath, add two or three boiling stones, followed by 2.5 grams of methyl salicylate. Place the flask in a sand bath and set up a macroscale reflux apparatus as shown by your lab instructor. Adjust the variac on the sand bath until the reaction mixture begins to boil and reflux it for 30 minutes.

2. Meanwhile prepare a 1M H_2SO_4 solution by diluting the stock solution found in the hood with deionized. Be careful, you are diluting a strong acid.

3. After 30 minutes cool the reaction mixture, pour it into a 250-ml beaker, and carefully add the 1M sulfuric acid until acidic to litmus. This may take as much as 75 ml of the acid. Then add an extra 10 ml of the 1M acid. The product should have precipitated.

4. Cool the reaction mixture in an ice bath and collect the product via vacuum filtration in a Buchner funnel. Rinse the product in the funnel with several portions of ice cold water.

5. Crystallize the crude product from hot water using a 125-ml Erlenmeyer flask. See experiment 4 for the proper technique. It may take as much as 50 ml of hot water to dissolve the solid. Gravity filter the hot solution through a preheated stemless funnel seated with fluted filter paper. Collect the filtrate in a clean 125-ml Erlenmeyer flask. Precipitation of the pure product should begin upon cooling. If not scratch the sides of the flask with a stirring rod to induce crystallization.

6. Cool the flask in an ice-bath for about 10 minutes then vacuum filter the product using a clean Buchner funnel. Rinse the salicylic acid with several portions of ice-cold water. Scrape the product into a watchglass and dry it in the oven for an hour.

7. Determine a yield and take a melting point.

8. In the data and results section calculate a theoretical and percent yield.

Name _____ Sec. _____ Exp. # _____

PROCEDURE	OBSERVATIONS

Name _____ Sec. _____ Exp. # _____

PROCEDURE	OBSERVATIONS

Name _____ Sec. _____ Exp. # _____

<u>DATA AND RESULTS</u>

Weight of product:

m.p. of product:

Write an equation for the reaction and calculate the percent yield:

QUESTIONS

1. Using curved arrow notation write a mechanism for the alkaline hydrolysis of methyl benzoate followed by acidification.

2. The product, salicylic acid, is soluble in hot water as you saw during crystallization. Draw the Lewis structure of the product and show the intermolecular interactions with water that permit this solubility. (Hint: see experiment 2).

EXPERIMENT 16

THE SYNTHESIS AND IDENTIFICATION OF COMMON NATURAL ESTERS

INTRODUCTION

The ester is a common functional group found in many naturally occurring organic compounds. Not only are they present in biological molecules called saponifiable lipids, but also they are responsible for the fragrances of perfumes and flowers. The odor and flavor components of common fruits such as oranges, bananas, pears, and pineapples are esters. In this experiment the class will prepare a series of esters and relate them to particular fragrances.

The easiest way of preparing an ester is to mix an alcohol with a reactive carboxylic acid derivative such as an acid chloride or anhydride.

A method producing lower yields due to the reversibility of the reaction involves heating a carboxylic acid and alcohol in the presence of a mineral acid catalyst.

In today's experiment a series of esters will be prepared for detection of odor. The products will not be isolated. Because yield is not an issue here, the method of heating a carboxylic acid with an appropriate alcohol in the presence of sulfuric acid will be used. Also you will not have to do a reactant table, although you will still have to look up the safety precautions utilized in handling these chemicals.

Name _____ Sec.: _____ Exp. #: _____

Safety/Hazards

PROCEDURE

1. You will work in pairs in this experiment. Your lab instructor will assign you a specific alcohol and carboxylic acid to use. These can be found in the hood. Also in the hood will be a series of boiling water baths.

2. Add about 1 ml of the assigned carboxylic acid to a large test tube. If the acid is a solid, use 1 gram. Add 5 ml of the assigned alcohol and mix. Then add 15 drops of concentrated sulfuric acid down the side of the test tube and mix again. If you get any sulfuric acid on your skin, wash the affected area thoroughly with water.

3. Place the reaction mixture in the water bath. Students with other assigned acids and alcohols will be sharing your water bath so be sure you label your test tube. After about 15 minutes remove the tube from the bath, and try to detect the fruity odor present in your test tube. If you do not detect any, heat the test tube for a longer period of time. The possible fragrances are banana, rum, peach, pear, orange, and wintergreen.

4. The possible carboxylic acid-alcohol mixtures are as follows:
   a. acetic acid and benzyl alcohol
   b. acetic acid and n-propyl alcohol
   c. acetic acid and 1-octanol
   d. acetic acid and isoamyl alcohol
   e. propionic acid and isobutyl alcohol
   f. salicylic acid and methyl alcohol

5. Using these alcohols and acids, complete Table 1 showing their structures as well as the structure of the expected ester. Name each ester and place the name in Table 2. Then complete the table by matching the appropriate fragrance with the ester. Because you were only assigned one, you need to find students who were assigned the others and use their test tubes for identifying the correct odor.

6. Using your textbook and/or lecture notes write the mechanism for the formation of your assigned ester in the space immediately below Table 2.

Name _____ Sec. _____ Exp. # _____

PROCEDURE	OBSERVATIONS

Name _____ Sec. _____

Table 1.

Carboxylic Acid	Alcohol	Ester

Table 2.

Name of the Ester	Fragrance

Mechanism:

EXPERIMENT 17

THE MICROSCALE HALOGENATION OF A BENZENE RING
I. Bromination of Acetanilide

INTRODUCTION

The halogenation of aromatic rings *via* electrophilic aromatic substitution is a common reaction performed in the organic chemistry laboratory. The bromination of a typical substrate such as acetanilide is shown below. The "acetamido" group already on the ring is an *ortho-para* director. As you learned in lecture, ferric bromide is a Lewis acid catalyst that polarizes the bromine bromine covalent bond.

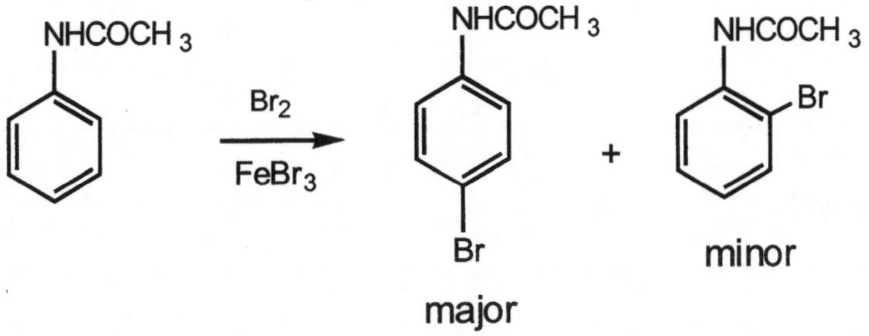

Liquid bromine used as a reagent in this reaction is very toxic and inconvenient to handle, and the small quantities needed for microscale work are difficult to measure precisely. Instead, we will be using pyridinium hydrobromide perbromide (also called pyridinium tribromide) to generate the Br_2 in situ. This is a fairly stable, crystalline solid that can be weighed accurately for microscale reactions.

Bromination occurs principally at the *para* position with a slight excess of the brominating reagent. Excess bromine remaining after ring substitution is complete can be removed by reduction with sodium hydrogen sulfite. The HSO_3^- is oxidized to SO_4^{-2} under acidic conditions. Ortho-substitution is very minor due to the steric hindrance exhibited by the large acetamido group. This amount is so small that crystallization of the product is not necessary.

Name _____ Sec. _____ Exp. _____

Reactant Table

Name of compound	m.p./b.p.	F.W.	Mass (g)	Vol. (ml)	D (g/ml)	moles

Safety/Hazards

PROCEDURE

1. Dissolve 340 mg of acetanilide in 3 ml of glacial acetic acid in the 10-ml Erlenmyer flask. Add 875 mg of pyridinium hydrobromide perbromide , swirl for several minutes, and place it in a water bath adjusted between the temperatures of 50°C to 60°C. Swirl occasionally and note your observations. Allow the reaction mixture to incubate for 30 minutes. Prepare about 5 ml of a saturated sodium bisulfite (sodium hydrogen sulfite) solution.

2. Remove the flask from the water bath and add 2 ml of water followed by about 40 drops of the saturated $NaHSO_3$ solution. Place the mixture into the water bath again. Although a precipitate will be present, the mixture should become colorless. If not, add more $NaHSO_3$ dropwise until the color dissipates.

3. Bring the mixture to room temperature and pour the contents into a 50-ml beaker containing 4 ml of water. Rinse the entire contents of the 10-ml Erlenmeyer into the beaker with several portions of water. Place the beaker in an ice-water bath for about 10 minutes as you set up a microscale vacuum filtration apparatus.

4. Stir the contents of the beaker for about 5 more minutes as it sits in the ice-water bath. Then vacuum filter the product into the Hirsch funnel. Rinse the product with several portions of ice-cold water, transfer to a pre-weighed watchglass, and place it in a 90°C oven for 15 minutes as you prepare to take a melting point.

5. Place a small amount of the product in a melting point capillary, but permit the watchglass to remain in the oven. After you are finished with the melting point, remove the watchglass from the oven, cool, and weigh.

6. Place your waste in the appropriate waste container in the hood.

Name_____ Sec. _____ Exp. # _____

PROCEDURE	OBSERVATIONS

DATA AND RESULTS

Weight of product:

m.p. :

Calculation of percent yield:

QUESTIONS

1. Write a mechanism for the synthesis that you just performed

2. Predict the major products if the following substrates were brominated. Draw the structures and provide the IUPAC names

a. Ethyl benzoate

b. *p*-methoxy ethyl benzoate

2. What does the final rinse of the product with ice-cold water remove?

EXPERIMENT 18

THE MICROSCALE HALOGENATION OF A BENZENE RING
II. Bromination of Vanillin

INTRODUCTION

The halogenation of aromatic rings via electrophilic aromatic substitution is a common reaction performed in the organic chemistry laboratory. The bromination of a typical substrate such as acetanilide is shown below. The "acetamido" group already on the ring is an *ortho-para* director. As you learned in lecture, ferric bromide is a Lewis acid catalyst that polarizes the bromine bromine covalent bond.

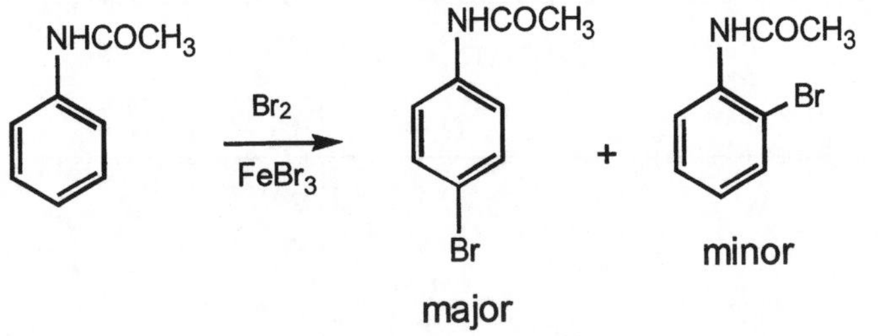

Liquid bromine used as a reagent in this reaction is very toxic and inconvenient to handle, and the small quantities needed for microscale work are difficult to measure precisely. Instead, we will be using pyridinium hydrobromide perbromide (also called pyridinium tribromide) to generate the Br_2 in situ. This is a fairly stable, crystalline solid that can be weighed accurately for microscale reactions.

In the case of acetanilide, bromination occurs principally at the *para*-position with a slight excess of the brominating reagent. (See the previous experiment). In this experiment you will be brominating vanillin. As you saw in lecture, the phenolic hydroxyl very highly activates the *ortho-para* positions. Because an aldehyde is already para to the phenolic –OH, substitution occurs at the *ortho* to the alcohol producing 5-bromo-4-hydroxy-3-methoxybenzaldehyde. Excess bromine remaining after ring substitution is complete can be removed by reduction with sodium hydrogen sulfite. The HSO_3^- is oxidized to SO_4^{-2} under acidic conditions.

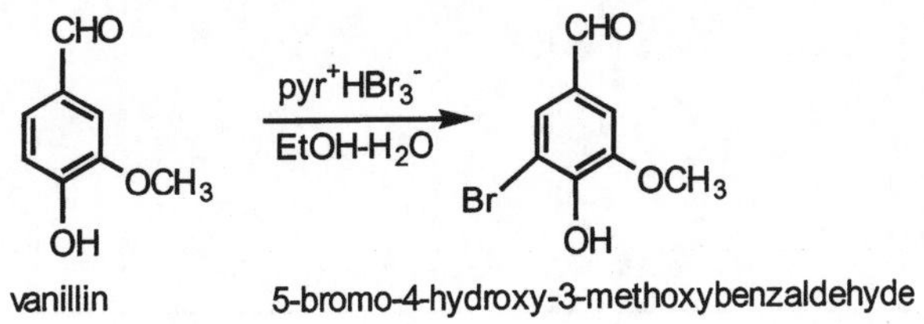

vanillin   5-bromo-4-hydroxy-3-methoxybenzaldehyde

Name _____ Sec. _____ Exp. # _____

Reactant Table

Name of compound	m.p./b.p.	F.W.	Mass (g)	Vol. (ml)	D (g/ml)	moles

Safety/Hazards

PROCEDURE

1. Dissolve 200 mg of vanillin in 3 ml of ethanol in a 25-ml Erlenmeyer flask. Add 340 mg of the pyridinium hydrobromide perbromide and swirl to dissolve. Place the flask in a water bath at 40°C to 50° C. Note the color of the solution.

2. Add water 1 milliliter at a time from a Pasteur pipet over a period of 10 minutes until the final volume is about 10 ml. The color should begin to dissipate until it is a light yellow color. Continue to heat it with occasional swirling for another 10 minutes.

3. Remove the reaction mixture from the water bath and add 1 ml of a saturated aqueous solution of sodium bisulfite. Then add another 5 milliliters of water and scratch the mixture with a stirring rod to induce crystallization. Cool the reaction mixture in an ice-water bath and continue to scratch for another 10 minutes.

4. Filter the crude product on a Hirsch funnel and rinse with three 1-ml portions of ice-cold water.

5. Perform a microscale recrystallization from methanol-water (see Experiment 4) and dry the product in the oven.

6. Weigh the product and record. Determine the percent yield in the data and results section.

7. Determine a melting point and record.

8. Place your waste in the appropriate container in the hood.

Name _____ Sec. _____ Exp. # _____

PROCEDURE	OBSERVATIONS

DATA AND RESULTS

Weight of product:

m.p. :

Calculation of percent yield:

Name ____ Sec. _______ Exp. # _____

QUESTIONS

1. What does the final rinse of the product with ice-cold water remove?

2. Write a mechanism for the synthesis that you just performed

3. Predict the major products if the following substrates were brominated. Draw the structures and provide the IUPAC names

a. Ethyl benzoate

b. *p*-methoxy ethyl benzoate

EXPERIMENT 19

THE MICROSCALE NITRATION OF ACETANILIDE

INTRODUCTION

In lecture you learned how to predict the position where electrophilic aromatic substitution occurs on aromatic rings that were already substituted. This process is based on knowing the directive influences of the substituents already on the ring. Except for the halogens, all ortho:para directors are activators and **accelerate** regioselective attack of the electrophile. In contrast, meta directors deactivate the ring and retard o:p substitution in favor of meta substitution.

Ortho:para directors				Meta directors
Very highly activates	Highly activates	Weakly activates	Weakly deactivates	Highly deactivates
-NH$_2$	-OCH$_3$	-R	-Br	-COOH
-OH	-NHCOCH3	-Ph	-Cl	-CHO
				-CN
				-NO$_2$

Looking at the structure of p-bromoacetanilide and using the table above, one would predict that electrophilic substitution will occur ortho to the acetamido group because it activates this position and the para position to it is already substituted. The electron withdrawing inductive effect of the bromo group deactivates its ortho position in favor of meta substitution.

In this experiment you will verify the prediction above by nitrating p-bromoacetanilide using a mixture of sulfuric and nitric acids. After isolating the product, its melting point can be compared to an authentic sample of 4-bromo-2-nitroacetanilide.

Name _____ Sec. _____ Exp. # _____

Reactant Table

Name of compound	m.p./b.p.	F.W.	Mass (g)	Vol. (ml)	D (g/ml)	moles

Safety/Hazards

PROCEDURE

1. Dissolve 1 millimole of p-bromoacetanilide in 3 ml of concentrated sulfuric acid in your 10-ml Erlenmyer flask. Place it in an ice bath.

2. In a small test tube carefully mix 20 drops of concentrated sulfuric acid and 15 drops of concentrated nitric acid. Let this mixture cool in an ice bath for 10 minutes.

3. Using a Pasteur pipet carefully add 1 ml (20 drops) of the H_2SO_4-HNO_3 mixture at a rate of about 1 drop every 30 seconds or so to the Erlenmeyer flask. Be sure to swirl the flask well after each addition.

4. After the addition is complete swirl the mixture for another 20 minutes as it is sitting in the ice bath.

5. Carefully pour the mixture into a small beaker containing 25 ml of ice. **Be careful**. You are adding a strong acid to water. Using a small stirring rod stir the mixture and scratch the sides of the beaker to induce crystallization.

6. Collect the precipitate on a Hirsch funnel and wash it with ice-cold water.

7. Dry the product in an oven and determine the yield.

8. Take a melting point of the product and calculate the percent yield in the data and results section

Name_____ Sec. _____ Exp. # _____

PROCEDURE	OBSERVATIONS

Name _____ Sec. _____ Exp. # _____

PROCEDURE	OBSERVATIONS

DATA AND RESULTS

Weight of product:

m.p. :

Write the equation for the reaction and calculate the percent yield:

QUESTIONS

1. Does the melting point of your product verify the prediction of where electrophilic aromatic substitution occurs? Explain.

2. Predict the major product of the following substitution reactions.

EXPERIMENT 20

THE MICROSCALE ACID-CATALYZED HYDROLYSIS OF AN AMIDE

INTRODUCTION

In this experiment you will explore the ability of amides to undergo hydrolysis under acidic conditions to produce a carboxylic acid and an amine. Note that in Experiment 17 acetanilide is brominated producing p-bromoacetanilide. And the previous experiment explored the nitration of the p-bromoacetanilide.

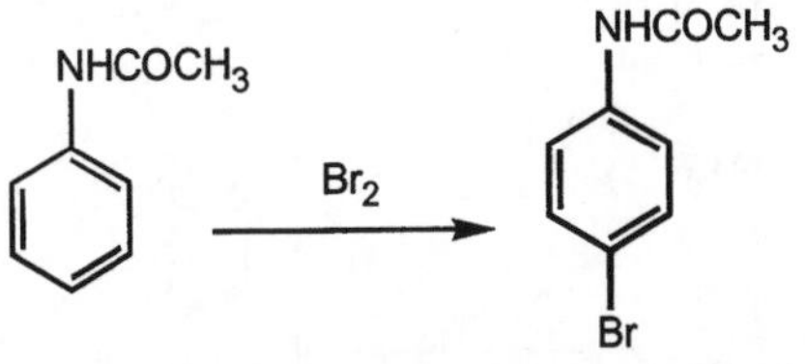

p-bromoacetanilide

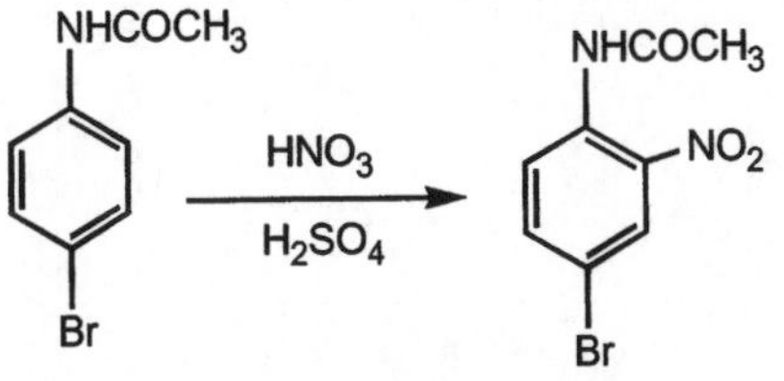

2-nitro-4-bromoacetanilide

In both cases the products have an "amide" functional group. Hydrolysis of the amide permits the organic chemist to access two new substituted anilines. Anilines are important precursor molecules in the synthetic dye industry.

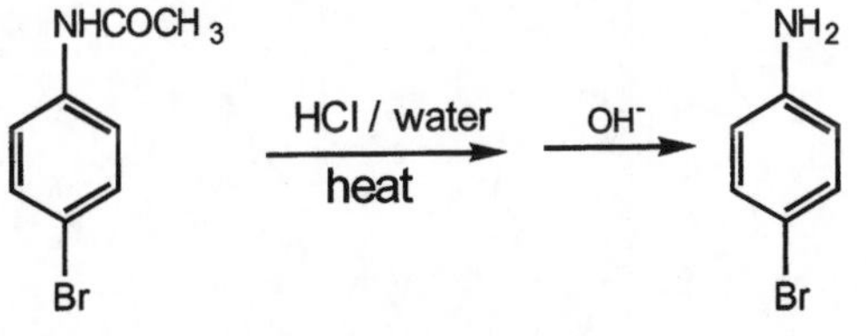

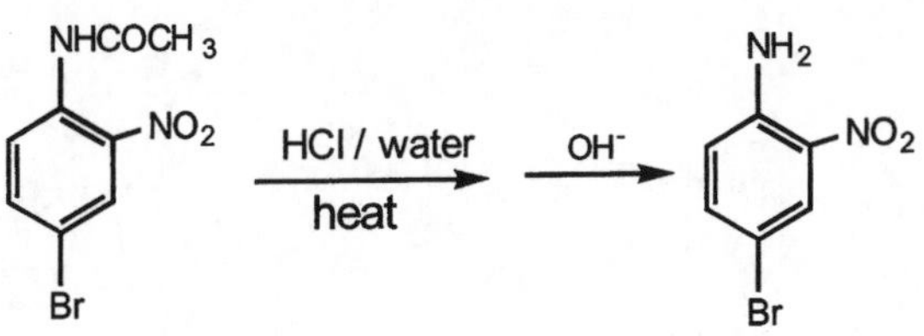

Wouldn't it be easier to prepare p-bromoaniline by simply brominating aniline? After all, The –NH$_2$ group is an ortho-para director. The problem is that the amino group so very highly activates the ring that bromination occurs at all of the ortho and para positions producing a tribromoaniline derivative.

2,4,6-tribromoaniline

This limitation is overcome by acetylating the amino group making it less of an activator and permitting momosubstitution to occur. The acetylation is normally performed using acetic anhydride (Ac$_2$O) and an acid catalyst. This protecting group can then be removed later as we are doing in this experiment.

Name _____ Sec. _____ Exp. # _____

Reactant Table

Name of compound	m.p./b.p.	F.W.	Mass (g)	Vol. (ml)	D (g/ml)	moles

Safety/Hazards

PROCEDURE

In this experiment you will work out your own procedure and write it up in the left-hand column of the procedure/observation pages. If you did the previous experiment, then use the 4-bromo-2-nitroacetanilide that was synthesized. Otherwise, begin with 1 to 2 millimoles of the 4-bromo-2-nitroacetanilide provided by your lab instructor.

1. You will need to prepare an HCl-H$_2$O solution that is between 6 to 8 Molar by diluting some concentrated HCl (12 M). Then do a gentle microscale reflux of your starting material with the aqueous acid. About 3 ml of acid for every millimole of substrate will do fine. Remember the technique for preventing evaporation as it boils?

2. Then the hydrolyzed product is poured into a mixture of ice and enough base to neutralize all of the HCl. You will need to do that calculation. Recall from previous experiments how you determine whether something is ultimately basic. You will need to choose a base. Ask your lab instructor if your choice is appropriate.

3. The product should precipitate. How will you collect it and what should the product be washed with to remove excess base and salts?

4. You will need to recrystallize the product. Using small amounts of your product test various procedures such as hot water, hot alcohol, or a two-solvent alcohol-water mixture.

5. Develop a way of determining whether you have the right product and whether it is pure or not. You may want to try TLC. Record all of this on the data and results page.

6. Calculate a percent yield.

Name _____ Sec. _____ Exp. # _____

PROCEDURE	OBSERVATIONS

Name _____ Sec. _____ Exp. # _____

PROCEDURE	OBSERVATIONS

Name _____ Sec. _____ Exp. # _____

PROCEDURE	OBSERVATIONS

DATA AND RESULTS

Weight of product:

Calculation of percent yield:

QUESTIONS

1. Write the equation showing the soluble acid hydrolyzed product that forms before the base is added.

2. Write the equation showing what happens when the product is poured into the cold solution of base.

EXPERIMENT 21

THE ISOMERIZATION OF MALEIC ACID TO FUMARIC ACID

INTRODUCTION

Below are the structures of two dicarboxylic acids: Their common names are maleic acid and fumaric acid. The ionized form of fumaric acid plays a major role in carbohydrate metabolism. It is one of the substrates in the Krebs cycle. Also notice that the two compounds are stereoisomers of one another since there is no free rotation around a carbon-carbon double bond. More specifically, they are often referred to as geometrical isomers. With this in mind, write the IUPAC names for both in the blank provided.

H₃C   CH₃
HOOC   COOH

Maleic Acid

H₃C   COOH
HOOC   CH₃

Fumaric Acid

_____   _____

In this experiment you will convert maleic acid to fumaric acid by heating it in HCl. In doing so, you will validate some of the principles learned in both lecture and lab. In preparing for your prelab quiz, think about the following questions:

1. What role does the acid play in this isomerization? What kind of intermediate forms? Hint: think about what happens between electrons in p-orbitals and H^+.

2. In preparing for this experiment you had to look up melting points. Why is the melting point of fumaric acid much higher than maleic acid. Hint: See the introduction to Experiment 1.

3. In this experiment you will find that maleic acid is many times more soluble in water than fumaric acid. Because "like dissolves like," maleic acid must be more polar than its stereoisomer. Why?

Name _____ Sec. _____ Exp. # _____

Reactant Table

Name of compound	m.p./b.p.	F.W.	Mass (g)	Vol. (ml)	D (g/ml)	moles

Safety/Hazards

PROCEDURE

1. Weigh 200 mg of maleic acid and transfer the solid to a 5-ml roundbottom flask. Attach a fractionating column to the flask, clamp the apparatus to a sand bath, and heat the mixture until the maleic acid dissolves. Carefully and slowly add 2 ml of concentrated HCl to the mixture through the fractionating column. Heat for another 10 minutes. You should see white crystals depositing on the side of the flask. Remove the reaction mixture from the sand bath and place it in lukewarm water for about 5 minutes. Then move it to an ice bath for 10 minutes to maximize the precipitation of product.

2. As the product is cooling, set up your microscale vacuum filtration system using the filtration flask and Hirsh funnel. Collect the product in the funnel and wash it with ice cold water. Describe what the crystals look like in the observation section.

3. The product melts too high for a melting point to be taken. Instead, various tests will be performed. In separate test tubes add 20 ml of DeI water to 100 mg of maleic acid, fumaric acid, and your product. Observe their abilities to dissolve in the water and record the results in Table1. Using portions of these solutions, do the other tests indicated in Table1, and record your results.

Table I.

Test	Maleic Acid	Fumaric Acid	Your Product
Solubility in water			
pH			
Reaction with magnesium			
Reaction with NaHCO$_3$			
Reaction with one drop of thymol blue			

Based on the results of these tests, is your product maleic or fumaric acid? Explain.

Name _____ Sec. _____ Exp. # _____

PROCEDURE	OBSERVATIONS

Name _____ Sec. _____ Exp. # _____

PROCEDURE	OBSERVATIONS

QUESTIONS

1. Write a mechanism for the isomerization of maleic acid to fumaric acid.

2. Below are the pK$_a$s of the two carboxyl groups in maleic and fumaric acids.

   Maleic
   pK$_{a1}$ = 1.8
   pK$_{a2}$ = 6.6

   Fumaric
   pK$_{a1}$ = 3.0
   pK$_{a2}$ = 4.5

   Why is the first carboxyl group of maleic acid more acidic than its isomer, but the second carboxyl group is less acidic? Hint: Look at the stability of the conjugate base after the first proton is removed.

EXPERIMENT 22

MICROSCALE PREPARATION OF TRIPHENYL

CARBINOL USING THE GRIGNARD REACTION

INTRODUCTION

The most important reactions to an organic chemist are those that involve carbon carbon bond formation. These permit the syntheses of larger, more complex molecules from simple precursors. A well-known reaction of this genre is an organometallic synthesis developed around 1905 by a French chemist named Victor Grignard. The "Grignard reagent" is prepared by mixing an alkyl halide with magnesium metal in the presence of an aprotic, donor solvent such as diethyl ether (Et_2O) or tetrahydrofuran (THF). In the alkyl halide, RX, the carbon is an electrophilic center. Upon reaction with magnesium, the carbon is converted to a nucleophilic center.

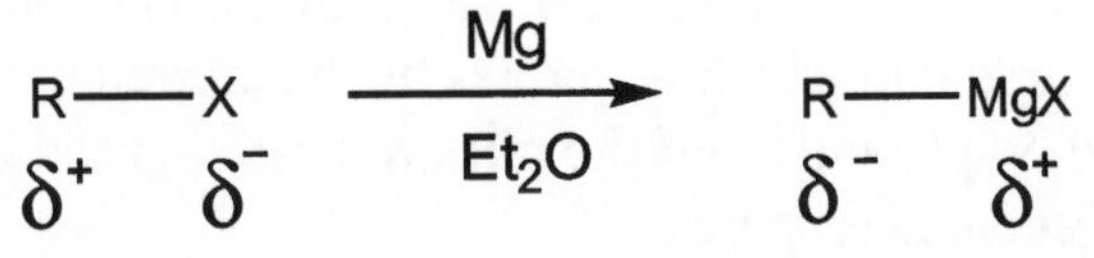

$$R\underset{\delta^+}{\text{———}}\underset{\delta^-}{X} \xrightarrow[Et_2O]{Mg} R\underset{\delta^-}{\text{———}}\underset{\delta^+}{MgX}$$

"Grignard Reagent"

"X" is typically Br or I. The donor solvent is essential for the formation of the Grignard reagent presumably by stabilizing it via Lewis acid Lewis base interactions as shown below.

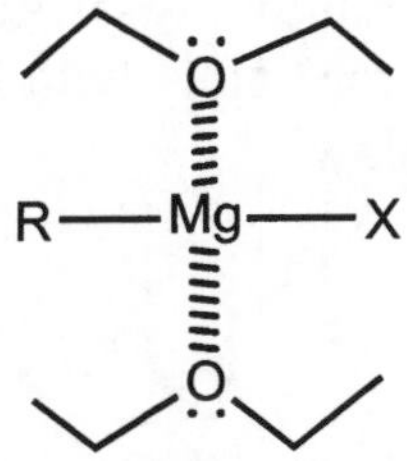

$$R\text{———}Mg\text{———}X$$

As you will see in lecture, this reagent will react with a myriad of organic substrates including ethylene oxide producing primary alcohols; aldehydes producing secondary alcohols; ketones and esters producing tertiary alcohols; and even carbon dioxide producing carboxylic acids.

In each case carbon carbon bond formation occurs. On the next page the generalized visualization of a Grignard reagent reacting with an aldehyde /ketone is shown. Note that the nucleophilic "R" group of the Grignard reagent has reacted with the electrophilic center of the carbonyl group producing a carbon carbon bond. The product is the conjugate base of the alcohol that is often referred to as the "salt of the alcohol." It exists simply because we are under basic conditions.

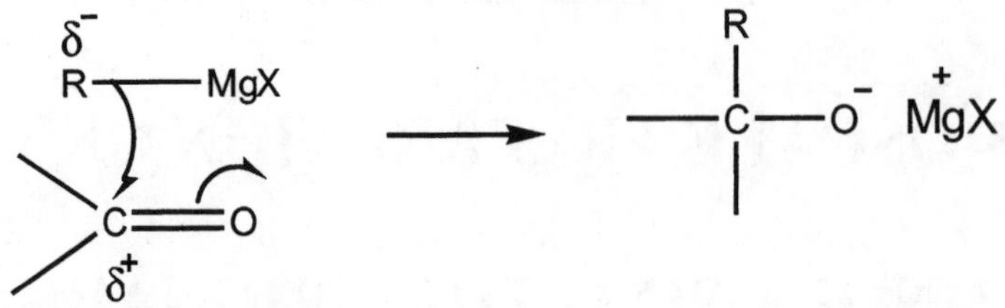

An aldehyde or ketone   salt of an alcohol

"Aqueous work-up" in dilute acid will protonate the conjugate base producing the alcohol.

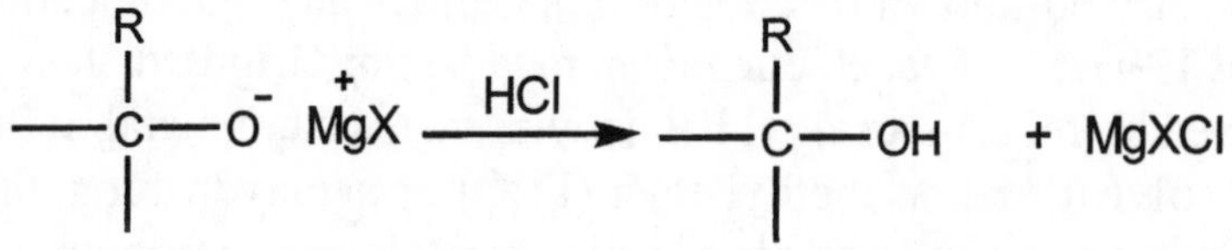

One limitation of the reaction is that the Grignard reagent is also a strong base and will undergo "protonolysis" with -OH, NH, SH, etc. groups. Thus the reagent must be prepared in an environment in which water has been scrupulously removed. This is usually accomplished by placing all of the glassware to be used in a 110-degree oven 30 minutes before the reagent is prepared.

Another limitation is that the magnesium metal must be fresh. Magnesium reacts with oxygen from the air upon standing and accumulates an oxide coating on its surface. Free magnesium metal is exposed by crushing the reagent with a spatula or in a mortar and pestle just before using.

In this experiment you will be generating the Grignard reagent using bromobenzene as the alkyl halide, then reacting it immediately with benzophenone producing triphenyl carbinol, a 3° alcohol, as shown below.

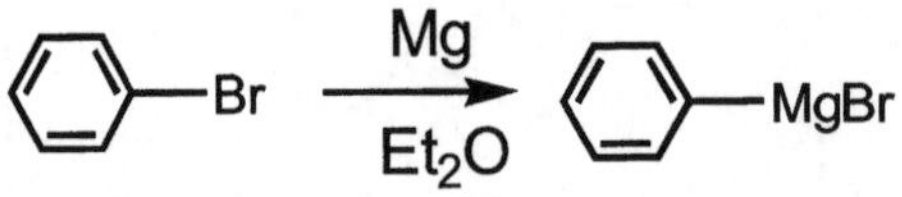

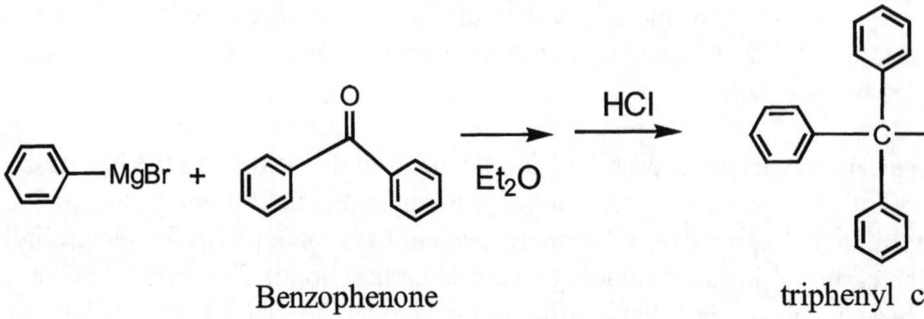

Benzophenone   triphenyl carbinol

Name _____ Sec. _____ Exp. # _____

Reactant Table

Name of compound	m.p./b.p.	F.W.	Mass (g)	Vol. (ml)	D (g/ml)	moles

Safety/Hazards

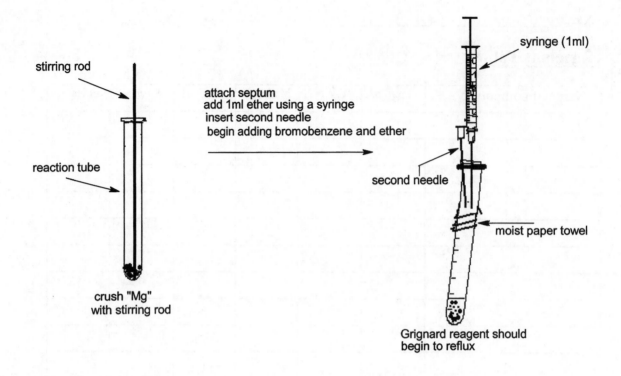

stirring rod

reaction tube

crush "Mg"
with stirring rod

attach septum
add 1ml ether using a syringe
insert second needle
begin adding bromobenzene and ether

syringe (1ml)

second needle

moist paper towel

Grignard reagent should
begin to reflux

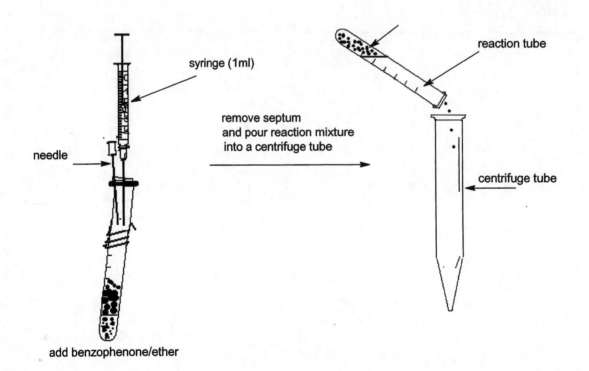

syringe (1ml)

needle

add benzophenone/ether

remove septum
and pour reaction mixture
into a centrifuge tube

reaction tube

centrifuge tube

186

PROCEDURE

1. For the Grignard reaction to work it is necessary that all of the equipment you use be dry. Before prelab lecture begins, place the following items in the oven: a microscale test tube, two septa, syringe with two needles, small stirring rod, and a glass vial. **Do not wash** any of these with water before drying. After prelab remove the items from the oven and place them in a desiccator over anhydrous calcium chloride.

2. Into the test tube weigh 75 mg of magnesium powder. Using the dry stirring rod crush the magnesium powder. Rinse the stirring rod with acetone and place it in the oven. You may need it later. Place a septum of the tube and with the syringe add 1 ml of diethyl ether. Then remove the vial from the desiccator, add 500 mg of bromobenzene, and dissolve in 1 ml of diethyl ether. Pull this solution into a syringe and insert it through the septum on the test tube. Rinse the vial with acetone and place it in the oven for later use.

3. Insert another needle (with no syringe) into the septum. This permits the release of air as you add things to the test tube. Then from the syringe add about 0.2 ml of the bromobenzene solution to the magnesium in ether and begin flicking the tube with your index finger. Within a minute or so the reaction should turn cloudy and begin to reflux. If not, remove the septum and crush the magnesium again with the stirring rod you re-dried in the oven until the reaction does start. Then add the remaining 0.8 ml of the bromobenzene dropwise over several minutes as you mix by flicking the tube. Make sure that the reaction mixture does not reflux so rapidly that the ether evaporates. If it does, wrap part of a wet paper towel around the tube. Finally, you need to remove the vial from the oven and place it in the desiccator.

4. The Grignard reagent you prepared is only stable for a short period of time. Remove the vial from the desiccator and quickly dissolve 550 mg of benzophenone in 1.0 ml of ether. Draw this solution into a dry syringe and insert into the septum of the tube containing the Grignard reagent. Add the benzophenone to the Grignard a 0.2 ml at a time with mixing. The reaction mixture should turn red. Again a moist paper towel can be used to control the reflux rate.

5. When finished pour the solution into your centrifuge tube. It may begin to solidify so this should be done quickly. You now have the magnesium salt of your product. You need to do aqueous acid work-up to protonate the salt. Add 3M HCl dropwise with stirring until the mixture is acidic to litmus. You will also need to add a milliliter or so of ether. You then have two fairly clear, light brown-colored layers. The triphenyl carbinol will be dissolved in the upper ether layer.

6. Using a Pasteur pipet, remove the aqueous layer and wash the ether layer with a saturated solution of NaCl. Remove the lower aqueous layer and dry the ether further using anhydrous $CaCl_2$ pellets.

7. Transfer the ether layer to a beaker and evaporate over a steam cone at very low heat in the hood. Add cold petroleum ether and stir the precipitate well. Then filter the product into a Hirsch funnel and rinse with a small portion of cold petroleum ether.

8. Remove a small amount to be dissolved in acetone later for spotting on a TLC plate. Then perform a microscale recrystallization of the product from a solvent assigned by your lab instructor.

9. Weigh the product and calculate a percent yield.

10. Determine the purity of your product on TLC. Your lab instructor will provide you with a choice of solvent systems. Also spot some benzophenone and crude product on your chromatogram before developing.

11. Run an IR spectrum of your product.

Name _____ Sec. _____ Exp. # _____

PROCEDURE	OBSERVATIONS

Name _____ Sec. _____ Exp. # _____

PROCEDURE	OBSERVATIONS

Name _____ Sec. _____ Exp. # _____

DATA AND RESULTS

Yield of Triphenyl carbinol (g) _____

Write the equations for the synthesis and calculate the theoretical and percent yields:

Sketch your chromatogram and calculate the R_f values for the spots:

Attach your IR to the experiment. List the prominent peaks observed in the IR.

QUESTIONS

1. Discuss the results of your TLC with respect to the purities of the crude product as well as the recrystallized product.

2. Write an equation showing what happens if the apparatus isn't dry and the Grignard reagent reacts with water.

3. Besides water, what is being removed from the ether layer when it is washed with a saturated NaCl solution?

4. What is the purposed of washing the crude product with petroleum ether before filtration?

EXPERIMENT 23

THE MICROSCALE SYNTHESIS OF BENZO-PINACALONE *VIA* A CARBOCATION REARRANGEMENT

Introduction

In lecture you saw that reactions in which the mechanism that proceeds through a carbocation intermediate can often undergo rearrangement. This can involve the migration of a hydride or an "R" group to a neighboring carbon. In this experiment you will verify this in the synthesis of benzopinacalone from benzopinacol.

benzopinacol   benzopinacalone

Experimental evidence suggests that the mechanism proceeds through a cyclic phenonium ion intermediate:

cyclic phenonium ion

PROCEDURE

1. Transfer 200 mg of benzopinacol to a 5-ml short-necked roundbottomed flask. Add 1.5 ml of glacial acetic acid and swirl for several minutes. Then add a crystal of iodine (I_2) and swirl for a few more minutes. Note any color changes. Attach your microscale column to the roundbottomed flask and secure the entire apparatus to a ringstand. Heat the mixture to boiling using a sandbath. Moisten a paper towel with ice water and wrap it around the column. Continue to reflux for ten minutes. Every minute or so re-moisten the towel with ice water and re-wrap the column. The mixture can be stirred by gently rotating the ringstand.

2. Raise the apparatus from the sandbath and let the mixture come slowly to room temperature. Pre-weigh a small watchglass and set up the microscale Hirsch funnel for doing a vacuum filtration. Note the color and texture of the product. Add 2 milliliters of absolute ethanol and stir the mixture well. Transfer the product to the Hirsch funnel. You will need to use several more milliliters of ethanol to accomplish this.
Rinse the solid by adding ethanol using a Pasteur pipet to the Hirsch funnel to remove any color from the product. Pull a vacuum for 5 more minutes to help remove some of the ethanol.

3. Transfer the product to the pre-weighed watchglass and place it in the oven for 20 minutes. Allow the watchglass to cool and determine the yield in grams. Determine the melting point of your product and do an IR using a nujol mull. Calculate the theoretical and percent yields.

4. In the data and results section, assess the purity of your product by discussing your melting point range as well as the IR peaks present or absent.

Name _____ Sec. _____

Reactant Table

Name of compound	m.p./b.p.	F.W.	Mass (g)	Vol. (ml)	D (g/ml)	moles

Safety/Hazards

Name _____ Sec _____

PROCEDURE	OBSERVATIONS

Name _____ Sec. _____

PROCEDURE	OBSERVATIONS

Name _____ Sec. _____

DATA AND RESULTS

Yield of benzopinacalone (g) _____

Theoretical yield (g) _____

Percent yield _____

Write the equation for this synthesis below and show all of your calculations:

QUESTIONS

1. Why would you expect the cyclic phenonium ion on page 1 to be especially stable? Use structures, if necessary.

2. It has been observed that the synthesis you just performed occurs at a much slower rate if there is a *nitro* group substituted on the benzene rings. Explain.

3. It has been proposed that the I_2 participates in the reaction by polarizing the OH, turning it into a good leaving group. With information as well as that found on page 1, write a detailed mechanism for the reaction you just performed in the laboratory.

4. What is the purpose of using the cold, wet paper towel in this experiment?

SECTION 3

THE CHEMISTRY OF NATURAL PRODUCTS

Cholesterol is a "natural product" that belongs to a major class of biological molecules called lipids. It is also one of a series of important molecules called steroids that are widely distributed in plants and animals. Steroids (I) have a common carbon skeleton consisting of four fused rings often labeled A,B,C, and D:

I

cholesterol

II

Cholesterol itself (II) also contains two important functional groups: an alkene between carbons 5 and 6 and an alcohol on carbon 3. It is soluble in the hydrophobic area of membranes and intimately involved in the fluidity of many animal membranes. It is present in animal fats, eggs, and other dairy products, and there is strong evidence that large concentrations of cholesterol in the blood plasma contribute to atherosclerosis (hardening of the arteries).

Cholesterol is the most abundant steroid found in the mammalian body. A 185-pound man contains about 280 grams of cholesterol distributed between the blood and all his tissue. It is also important because mammalian cells use cholesterol to biosynthesize all of our important steroid hormones such as progesterone, testosterone, and corticosterone. All of these have the A,B,C,D fused ring skeleton, but differ in functional groups that are attached to the steroid nucleus. Therefore, having an understanding of the reactivity of cholesterol is important. We will examine this reactivity in several experiments.

The cyclohexane rings in steroids are in a chair conformation. Actually they adopt a *trans*-fused conformation in which the rings are rigid and relatively flat because they cannot undergo a chair flip as typical cyclohexane rings do. The two methyl groups on the steroid are in axial positions and are often referred to as "angular methyls." The top face of the ring where the angular methyl groups are is referred to as the β-face while underneath the ring is called the α-face. This structure (III) can be found on the next page. The alcohol group and isooctyl sidechain are also on the β-face as can be seen in structure (IV). This is also incorporated into the nomenclature system for steroids. For example, cholesterol is 5-cholesten-3β-ol indicating that the double bond is between

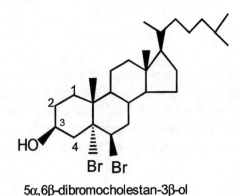

β-face

angular methyls   CH₃

CH₃

α-face   H   III

cholesterol

IV

carbons 5 and 6, and the alcohol is in the axial position. Later you will react cholesterol with bromine producing the trans-dibromide because bromination is an anti-addition. Thus one bromine is on the β-face while the other is on the α-face. This is also incorporated into the name as shown below (V).

HO

Br Br

5α,6β-dibromocholestan-3β-ol

V

EXPERIMENT 24

MICROSCALE PURIFICATION OF CHOLESTEROL

INTRODUCTION

At this point in organic chemistry you have performed a number of microscale reactions. The procedures were designed for you by downsizing known macroscale procedures. In this experiment you will design your own microscale experiment by writing a procedure from a published experiment. You will begin by obtaining a macroscale procedure from your college library. It can be found on pages 5421-5422 in volume 75 of the *Journal of the American Chemical Society* published in 1953. Note that it is a macroscale two-step procedure involving the bromination of the double bond of cholesterol followed by debromination with zinc dust producing pure cholesterol. The reactions involved can be written in the space below. You need to modify the experiment so that it can be performed on 1 millimole of crude cholesterol. The procedure should be handwritten in pencil on the left side of the procedure/observation pages. Your lab instructor will check your procedure when you arrive at lab and may suggest a few modifications to make the procedure a little safer.

REACTION SEQUENCE

Name _____ Sec. _____ Exp. # _____

Reactant Table

Name of compound	m.p./b.p.	F.W.	Mass (g)	Vol. (ml)	D (g/ml)	moles

Safety/Hazards

Name _____ Sec. _____ Exp. # _____

PROCEDURE	OBSERVATIONS

Name _____ Sec. _____ Exp. # ____

PROCEDURE	OBSERVATIONS

Name _____ Sec. _____ Exp. # _____

PROCEDURE	OBSERVATIONS

Name _____ Sec. _____ Exp. # ____

PROCEDURE	OBSERVATIONS

Name _____ Sec. _____ Exp. # _____

DATA AND RESULTS

Yield of pure cholesterol _____

m.p. of cholesterol _____

EXPERIMENT 25

MICROSCALE BROMINATION OF CHOLESTANONE

INTRODUCTION

A number of steroids are sex hormones that contain a ketone at the 3-position of the steroid nucleus. Examples are testosterone (I), the primary male sex hormone, and progesterone(II), a female sex hormone that prepares the uterus for the implantation of a fertilized egg.

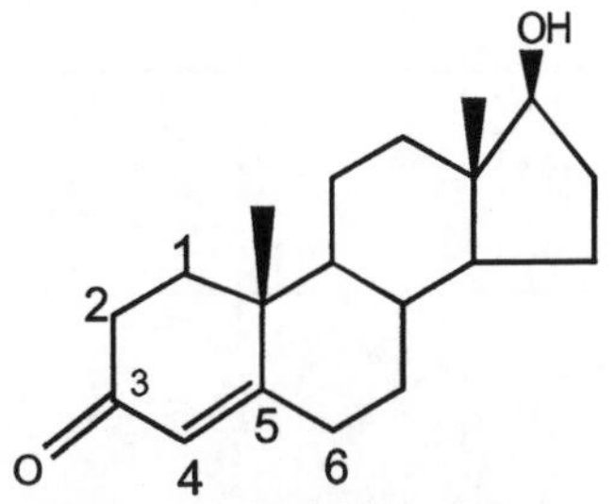

testosterone

I

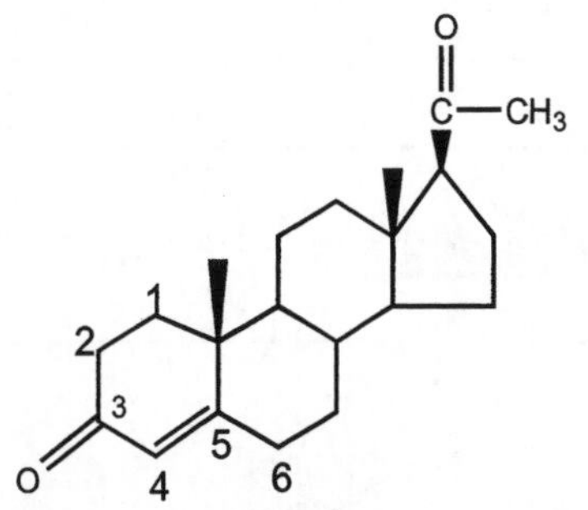

progesterone

II

As discussed in lecture, the α-position of a ketone is weakly acidic and is susceptible to bromination via an enolate anion (under basic conditions) or an enol intermediate (under acidic conditions). In this experiment you will see that the α-carbon of a steroid ketone can be stereoselectively and regioselectively brominated with pyridinium hydrobromide perbromide (acidic conditions). The substrate is 5α-cholestan-3-one (III) and the product 2α-bromo-5α-cholestan-3-one (IV). Write a mechanism for this reaction on the last page.

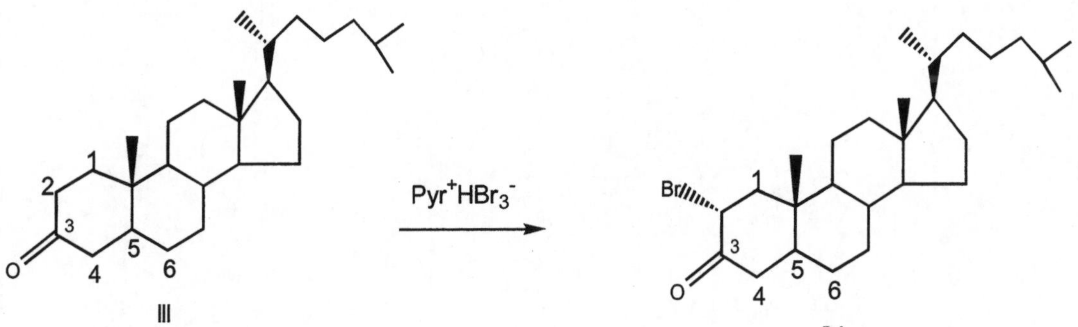

$Pyr^+HBr_3^-$

III

IV

Name _____ Sec. _____ Exp. # _____

Reactant Table

Name of compound	m.p./b.p.	F.W.	Mass (g)	Vol. (ml)	D (g/ml)	moles

Safety/Hazards

PROCEDURE

1. Using a 10-ml Erlenmeyer flask dissolve 100 mg of 5α-cholestan-3-one in 5 ml of absolute ethanol in a water bath at 50°C to 60°C. Add 80 mg of pyridinium hydrobromide perbromide and swirl to dissolve. Record the color and again place the reaction mixture in the water bath.

2. Let the flask remain in the water bath for about 30 minutes and occasionally swirl. Observe and record. The product should begin to precipitate as the reaction progresses.

3. Let the reaction flask come to room temperature and vacuum filter the product into a Hirsch funnel. Wash the product with several 1-ml portions of cold absolute ethanol. You should obtain a white, crystalline product.

4. Dry the product in the oven and determine a yield

5. Calculate a percent yield in the data and results section.

6. Determine a melting point for the product (IV).

Name _____ Sec. _____ Exp. # _____

PROCEDURE	OBSERVATIONS

Name _____ Sec. _____ Exp. # _____

DATA AND RESULTS

Yield of product (g) _____

Melting point of product (°C)

Theoretical yield (g) _____

Percent yield _____

Write the equation for the synthesis below and show all of your calculations:

Name _____ Sec. _____ Exp. # _____

QUESTIONS

1. What does the melting point tell you about the purity of your product?

2. Write a mechanism for the reaction using curved arrow notation

EXPERIMENT 26

MICROSCALE ESTERIFICATION OF CHOLESTEROL

INTRODUCTION

The alcohol group on the 3-position of cholesterol is susceptible to esterification. Cholesterol ester is the storage form of cholesterol in cells. It is biosynthesized from cholesterol and an acyl-coenzyme A. The synthesis is catalyzed by the enzyme, acyl-CoA:cholesterol acyl transferase (ACAT). The enzyme is located on the cytosolic surface of liver endoplasmic reticulum.

It is also possible to prepare cholesterol esters in the laboratory by heating cholesterol with an organic acid chloride. In this experiment you will prepare the benzoate ester of cholesterol. At the option of your lab instructor, you may be assigned a fatty acid chloride instead.

The pyridine is used to neutralize the HCl that is liberated as the reaction proceeds. The product is precipitated by adding methanol and can be recrystallized from ethyl acetate. Preparing the ester from a longer-chained fatty acid acyl chloride may require a different recrystallization solvent.

Name _____ Sec. _____ Exp. # _____

Reactant Table

Name of compound	m.p./b.p.	F.W.	Mass (g)	Vol. (ml)	D (g/ml)	moles

Safety/Hazards

PROCEDURE

1. The following procedure should be performed in a hood. To 1 millimole of cholesterol in a 10-ml Erlenmyer flask add 1 ml of pyridine. Swirl to dissolve the cholesterol and add 0.2 ml (your lab instructor will specify the correct number of drops) of benzoyl chloride from a Pasteur pipet. Heat the flask on a steam cone in the hood for about 15 minutes.

2. Cool the reaction mixture and add 5 ml of methanol to precipitate the product. You may need to induce crystallization by cooling and/or scratching. Vacuum filter the product into the microscale Hirsch funnel and wash it with three 1-ml portions of cold methanol.

3. Do a microscale recrystallization from ethyl acetate (see experiment 4) and again collect the product on a Hirsch funnel.

4. Dry the product for 10 minutes in an oven and determine a yield.

5. Do the melting point of the product and perform an IR to determine the absence of the alcohol stretching frequency.

6. Calculate a percent yield.

PROCEDURE	OBSERVATIONS

Name _____ Sec. _____ Exp. # _____

DATA AND RESULTS

Yield of the ester of cholesterol _____

m.p. _____

(attach the IR to the back of this page. Be sure your name is on the IR)

Write the equation for this reaction and calculate the theoretical and percent yields:

QUESTIONS

1. Does your IR suggest that the product is the ester rather than an alcohol? Explain

2. Write the equation showing the purpose of the pyridine in the reaction.

3. Is the product more or less polar than the starting material (cholesterol)? Explain.
   Hint: see experiment 2.

SECTION 4

STUDIES ON THE MECHANISTIC ASPECTS OF CHEMICAL REACTIONS

In this section of the lab manual we will study the mechanisms of a few simple organic reactions. A mechanism is a step-by-step process showing how bond-breaking and bond-making occurs for a specific chemical reaction. This goes hand-in-hand for some of the syntheses you have been doing in lab. Reaching an understanding of why a particular mechanism operates in any given situation enables one to develop an appreciation of the synthetic methods used in the production of organic compounds. Not only are mechanisms essential to understanding chemical processes, but also mechanistic theory can be used to explain diverse chemical phenomena in terms of a small number of basic principles. This makes it easier to understand organic chemistry as a whole. This is really why so much time is spent on writing mechanisms in the lecture section of sophomore organic chemistry. In this section we will examine the experimental evidence upon which mechanistic theory is built.

A few organic reactions occur in one step. These are said to be **concerted**. Most reactions occur in a series of two or more steps. There are various ways of deducing mechanistic pathways. The most common method involves studying the kinetics of a particular reaction. In the following experiments you will examine the mechanisms of several nucleophilic substitution reactions by measuring or estimating the rates reactions. Keep in mind that these results can help us understand other reactions. For example, you will study the rate of a reaction in which the kinetically slow step is the formation of a carbocation. The results of this experiment will demonstrate that 3° carbocations are more stable than 2° carbocations. This basic fact can then be used to understand how and why a variety of other reactions occur as they do.

EXPERIMENT 27

A KINETIC STUDY:

RATE EFFECT OF SOLVENT AND TEMPERATURE ON AN SN1 REACTION

INTRODUCTION

Learning organic chemistry involves understanding the relationships between the structure of molecules and how they function in reactions. As you often see in lecture, this requires that reactions be organized into categories and a mechanistic pathway be developed for each of these classifications. A mechanism can sometimes be determined by studying the kinetics of the particular chemical reaction. For example, in lecture you saw that a nucleophilic substitution reaction of an alkyl halide can be classified as being SN1 or SN2. Kinetics can be used to support the proposed mechanisms for these reactions.

The substitution of a nucleophile for the halogen of a primary alkyl halide occurs *via* a concerted process. This one-step SN2 mechanism can be verified by studying the rate of the reaction. It should be bimolecular: first order in nucleophile (Nu:) and first order in RCH_2Br.

$$RCH_2Br + Nu: \quad \rightarrow \quad RCH_2Nu$$

$$Rate = k\ [RCH_2Br]\ [Nu:]$$

This can be accomplished in the laboratory by measuring the rate of the reaction under specified concentrations of alkyl halide and nucleophile. If the proposed concerted mechanism is correct, then doing other experiments in which either RCH_2Br or Nu: are doubled should proceed at twice the rate of the first one.

When tertiary alkyl halides react with nucleophiles the SN1 mechanism predominates. It is not concerted, but instead, occurs in more than one step. In the first step, the alkyl halide loses its halogen to form a slightly stable carbocation. The nucleophile then attacks the carbocation in a second step. The first step is much slower than the first one and is thus rate determining. Because the tertiary alkyl halide is the only reactant in the slow step, the reaction should be unimolecular and follow the rate law below.

$$R_3CBr \quad \rightarrow \quad R_3C+ \quad + \quad Br^- \quad (slow)$$

$$R_3C+ \quad + \quad Nu: \rightarrow \quad R_3CNu \quad (fast)$$

$$Rate = k\ [R_3CBr]$$

The unimolecularity and proposed mechanism for this reaction is supported by kinetic measurements. Doubling the concentration of the tertiary alkyl halide doubles the reaction rate. While increasing the concentration of nucleophile has no effect on the speed of the reaction.

A typical SN1 reaction is the solvolysis of t-butyl chloride in water or alcohol.

Normally the difficulty encountered in doing chemical kinetics is finding a way to experimentally follow the disappearance of reactant or appearance of product. If we were to monitor the reaction above by following the disappearance of t-buCl, then the rate law would be

$$\frac{- \Delta \, (\text{t-buCl})}{\Delta \, \text{time}} = k \, [\text{t-buCl}]$$

The rate (change in concentration of t-buCl per unit time) equals the rate constant, k, times the remaining concentration of t-buCl at any instant. At zero time, the equation provides the initial rate. The value of a rate constant is useful for determining reaction conditions. The larger k is, the faster the reaction. Changing temperature or solvent will alter k. This can often be valuable in determining the mechanism of a reaction. Since the rate itself is always changing, it is easier to find k by using the "integrated rate law." Integration of the rate law above provides the following integrated rate law:

$$\text{Ln} \, \frac{(\text{t-buCl})_{\text{initial}}}{(\text{t-buCl})_t} = kt \quad (\text{where } t= \text{time in sec. or min.})$$

Thus, it is only necessary to precisely measure the time required for a certain percentage of the initial t-buCl concentration to react. For example, after 20% of the t-buCl reacts, the remaining t-buCl would be 80% (0.8). From this k can be calculated:

$$\text{Ln} \, \frac{1}{0.8} = kt$$

$$k = \frac{0.223}{t}$$

As you saw in lecture the energy barrier to any chemical reaction is called the activation energy (E_a). This activation energy is related to the rate constant (k) by the Arrhenius equation:

$$k = Ae^{-Ea/RT}$$

where T= Kelvin temperature and A is a constant and R is the ideal gas law constant.

Taking the natural logarithm of both sides puts it in the form of the equation of a straight line:

$$\ln k = -E_a/RT + \ln A$$

One can then determine activation energy by measuring the rate constant for a series of experiments at varying temperatures and plotting the ln k versus 1/T. The E_a can then be calculated from the slope of the resulting straight line.

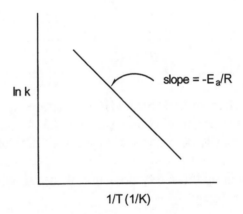

In this experiment you will study the effect of solvent polarity on the rate of the typical SN1 hydrolysis of t-butyl chloride.

CH₃—C(CH₃)(CH₃)—Cl   + ROH   →   CH₃—C(CH₃)(CH₃)—OR   + HCl

T-butyl chloride
(t-buCl)

R=H or Et

Note that the reaction generates HCl. Therefore, the rate can be followed by measuring the time required for the HCl to neutralize NaOH in the presence of a phenolphthalein indicator. By using the same amount of NaOH in each experiment, the relative rates can be determined. This enables us to determine what solvent system is better; however, it is not possible to calculate a rate constant. This would require that we precisely know the initial concentrations of t-butyl chloride and sodium hydroxide.

By choosing a specific solvent and measuring the rates at varying temperatures, the activation energy can be calculated from an Arrhenius plot. We are unable to calculate a rate constant, k, however, recall that k is inversely proportional to the time required for a certain percentage of a reactant to disappear (or a reactant to appear). Therefore, a plot of Ln time versus 1/T provides a straight line where the slope = + E_a/R.

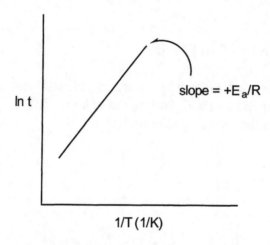

It is easy to measure the time required for the HCl generated to react with the same amount of NaOH at each temperature. If one uses R = 8.314 J/mol-K, then E_a will be in Joules. Using R=1.99 cal/mol-K permits us to express the activation energy in calories.

Before you come to lab look up the safety precautions that should be used and record those in the table on the next page.

Name _____ Sec. _____ Exp. _____

Safety/Hazards

PROCEDURE

Effect of Solvent on the SN1 Solvolysis of *tert*-butyl chloride.

1. Students will work in pairs. You will prepare mixtures of ethanol and water as well as another solvent (assigned by your lab instructor) and water. The composition of each mixture is given in Table 1.

2. Label four test tubes. To each test tube add 3.0 ml of the solvent system and five drops of 2% NaOH containing phenolphthalein indicator. Then add 1 drop of phenolphthalein indicator. Cover the tubes with parafilm and place them in a water bath at 30°C. After 5 minutes add five drops of t-butyl chloride to each tube and mix well. Note the time of addition and in Table 1 record the times required for the color to dissipate. Answer the questions with respect to the results of your experiment.

Effect of Temperature and Determination of Activation Energy

1. From your results in part A choose a solvent system whose reaction time was about 5 minutes at 30°C. Then repeat the experiment above at room temperature and at 35 and 40°C centigrade. Record the results in Table 2.

2. Plot ln (reaction time) on the y-axis and the reciprocal of temperature (Kelvin) on the x-axis for the four data points in the experiment. Have the program calculate the slope of the line.

3. Calculate the activation energy in both Joules and calories and record the results. Attach the graph to the experiment.

Name _____ Sec. _____ Exp. _____

DATA AND RESULTS

Table 1.

Solvent system	Time (sec)	Solvent system	Time (sec)
50:50 ethanol-water		50:50	
55:45 ethanol-water		55:45	
60:40 ethanol-water		60:40	
65:35 ethanol-water		65:35	

Table 2.

Solvent chosen:	
Temperature (^{o}C)	Time (sec.)

Calculation of activation energy.

QUESTIONS

1. Based on the results of part A, list the three solvents (ethanol, water, and the assigned solvent) in order of increasing solvolytic power. Is this the order (of polarity) you would expect based on the mechanism of the SN1 reaction? Explain.

2. In which ethanol-water mixture does the reaction have the larger rate constant? Explain.

3. The concentration of sodium hydroxide in this experiment is approximately 2%. Why isn't it necessary to know the precise concentration of the base?

4. Note that the activation energy is constant and does not change with temperature. What does the activation energy depend on? Explain

EXPERIMENT 28

THE SN2 REACTIVITY OF ALKYL HALIDES

INTRODUCTION

The halogens when bonded to specifically hybridized carbon are susceptible to replacement by a variety of nucleophiles. In this experiment you will study one of the types of substitution reactions as generalized below.

$$RX + Nu: \rightarrow RNu + X^- \text{ where X can be Cl, Br, or I}$$

The carbon to which the leaving group is attached is called the α-carbon. Kinetic evidence has shown that when RX is a primary or secondary alkyl halide, the mechanism is concerted and the reaction is second order. An example of this type of reaction is the displacement of bromine from an alkyl halide by the nucleophile, cyanide ion, in the solvent acetone:

Overall reaction:

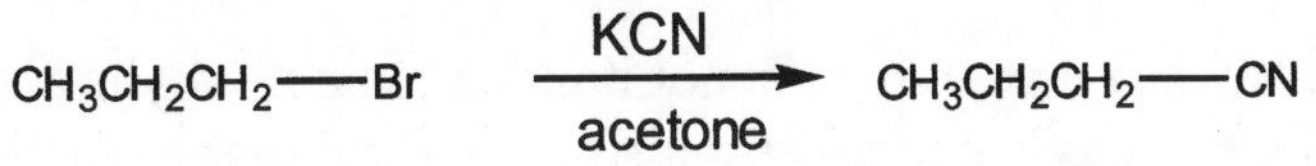

$$CH_3CH_2CH_2—Br \xrightarrow[\text{acetone}]{\text{KCN}} CH_3CH_2CH_2—CN$$

Mechanism:

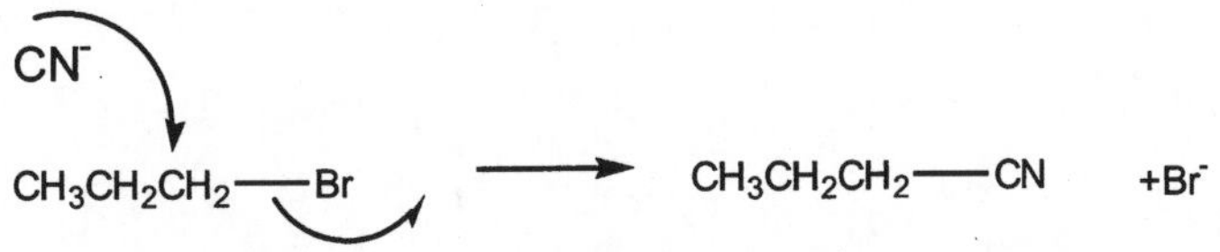

$$CN^- \quad CH_3CH_2CH_2—Br \longrightarrow CH_3CH_2CH_2—CN \quad +Br^-$$

The mechanism is concerted with the n-propyl bromide and cyanide ion colliding in the one and only step. Thus the reaction is first order in each reactant (second order overall) with the rate law:

Rate = k[n-propyl bromide][KCN]

This is typically referred to as an SN2 reaction where S=substitution, N=nucleophilic, and 2=second order. Experiments with optically active substrates demonstrate that the nucleophile approaches the α-carbon on the side opposite from which the leaving group departs. This leads to inversion of configuration. The typical SN2 reagent, sodium iodide (in acetone), can be used to study this stereoselectivity as shown in the equation on the next page.

CH₂CH₃ structure — (S)-sec-butyl chloride → NaI / acetone → (R)-sec-butyl iodide

Nucleophilic substitution of the substrate (S)-*sec*-butyl chloride leads to the production of (R)-*sec*-butyl iodide. This backside attack leading to inversion should be affected by Van Der Waals repulsions between substituents on the α-carbon and the approaching nucleophile. Thus the speed of SN2 reactions for secondary and tertiary alkyl halides should be slower than their primary counterparts. In this experiment you will test this by observing the rates of reaction of various alkyl halides.

Sodium iodide is used in the experiment for several reasons. It is a very powerful nucleophile, but yet is a weak base. Therefore it will not participate in alternate elimination reactions. It is soluble in acetone. Aprotic solvents such as acetone and dimethylformamide (DMF) are useful in SN2 reactions because they are slightly polar, yet are aprotic. This minimizes the ability of the solvent to solvolyze the nucleophile leading to diminished nucleophilicity. Also the product of the reactions are NaBr or NaCl, both of which are insoluble in acetone. This permits us to follow the course of the reaction by observing the formation of a precipitate.

Do SN2 reactions occur when the halogen is not bonded to an sp³-hybridized carbon? You will determine this in today's experiment also by observing the reactivity of bromobenzene with sodium iodide in acetone.

Before you come to lab look up the safety precautions that should be used and record these in the table on the next page.

SAFETY/HAZARDS

PROCEDURE

1. Clean a series of small test tubes by rinsing with acetone and drying in the hood with a stream of air. Do not use soap and water. The alkyl halides you will be using are not water miscible.

2. Into individual test tubes place five drops of the following alkyl halides: #1 bromobenzene, #2 tert-butyl chloride, #3 2-chlorobutane, #4 2-bromobutane, and #5 1-chlorobutane. Then to tube #1 add 40 drops of the sodium iodide-acetone reagent and mix well. Note the time of addition and record the seconds required for a precipitate to appear. If a precipitate does not form, set test tube #1 aside and repeat the same procedure for the remaining four tubes.

3. Several of the reaction tubes may not have produced a precipitate If so, place these in a water bath at 45°C for another 10 minutes and record the time required for precipitation to occur. Note those tubes that do not produce a precipitate.

4. Record your results in Table 1 in the data and results section. Calculate the elapsed time (if any) and record this in Table 2 along with the structure of the appropriate alkyl halide.

Name _____ Sec. _____

DATA AND RESULTS

Table 1.

Name of the Alkyl Halide	Initial time (sec.)	Precipitation (sec.)	Elapsed time (sec.)
#1			
#2			
#3			
#4			
#5			

Table 2.

Structure of the Alkyl Halide	Elapsed time (sec.)
#1	
#2	
#3	
#4	
#5	

Name _____ Sec. _____

Place the five alkyl halides you tested in order of decreasing reactivity with sodium iodide in acetone. Discuss in detail how this order is related to the structure of the individual substrates and the mechanism of the SN2 reaction.

QUESTIONS

1. Based on what you learned in this experiment, place the following alkyl halides in order of decreasing reactivity in SN2 reactions: vinyl chloride (chloroethene); ethyl bromide; cyclohexyl chloride; and cyclohexyl bromide.

2. Draw the structure of (R)-1-bromo-1-phenylpropane and predict the SN2 product if it were to react with NaCN in DMF.

3. Would you expect 1-bromo-1-methylcyclopentane to react with NaI/acetone? Explain.

EXPERIMENT 29

THE SN1 REACTIVITY OF SEVERAL ALKYL HALIDES

INTRODUCTION

By performing the previous experiment, you probably noticed that tertiary alkyl halides are unreactive in SN2 reactions. However, chemists observed that 3° substrates do undergo solvolysis reactions if placed in weakly nucleophilic, polar solvents such as water or alcohol. As usual, the mechanistic pathway of this reaction was determined by studying the kinetics of the reaction. The reaction was determined to be unimolecular and proceeds via the SN1 mechanism where S=substitution, N=nucleophilic, and 1=unimolecular (first order). It was observed that doubling the concentration of the tertiary alkyl halide doubles the reaction rate while increasing the concentration of nucleophile has no effect on the speed of the reaction.

The reaction is not concerted, but instead, occurs in more than one step. In the first step, the alkyl halide loses its halogen to form a slightly stable carbocation. The nucleophile then attacks the carbocation in a second step. The first step is much slower than the first one and is thus rate determining. Because the tertiary alkyl halide is the only reactant in the slow step, the reaction should be unimolecular and follow the rate law below.

$$R_3CBr \rightarrow R_3C+ + Br^- \quad (slow)$$

$$R_3C+ + Nu: \rightarrow R_3CNu \quad (fast)$$

$$Rate = k\,[R_3CBr]$$

A typical SN1 reaction is the solvolysis of t-butyl chloride in water or alcohol.

T-butyl chloride
(t-buCl)

If an SN1 reaction proceeds via an sp^2-hybridized, carbocation intermediate then racemization of α-carbon stereocenters should be observed as diagrammed on the next page. Experiments have shown that this does occur.

241

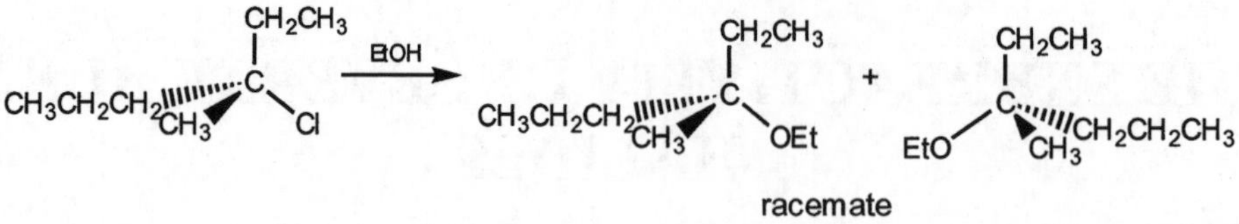

CH$_2$CH$_3$

CH$_3$CH$_2$CH$_2$—C—Cl $\xrightarrow{\text{EtOH}}$ CH$_3$CH$_2$CH$_2$—C—OEt   +   EtO—C—CH$_2$CH$_2$CH$_3$
          | | |
         CH$_3$ CH$_3$ CH$_3$
                                                  CH$_2$CH$_3$ CH$_2$CH$_3$

racemate

The rate of the SN1 reaction should increase as stability of the carbocation intermediate increases. You will verify the relative trend of carbocation stability in this experiment by determining the relative SN1 solvolysis reactivity of 1°, 2°, and 3° alkyl halides.

The reagent used in this experiment is silver nitrate in ethanol. This permits you to follow the reaction since alcohol insoluble silver halide is produced as solvolysis occurs. Also, the silver ion presumably accelerates the SN1 reaction by weakly complexing to the halogen leaving group thus weakening its bond to carbon:

$$R----X----Ag^+$$

Before you come to lab look up the safety precautions for all of the chemicals being used in this experiment and record these in the table on the next page.

Name _____ Sec. _____

SAFETY/HAZARDS

PROCEDURE

1. Clean a series of small test tubes by rinsing with acetone and drying in the hood with a stream of air. Do not use soap and water. The alkyl halides you will be using are not water miscible.

2. Into individual test tubes place five drops of the following alkyl halides: #1 bromobenzene, #2 tert-butyl chloride, #3 2-chlorobutane, #4 2-bromobutane, #5 1-chlorobutane. Then to tube #1 add 40 drops of the ethanolic silver nitrate reagent and mix well. Note the time of addition and record the seconds required for a precipitate to appear. If a precipitate does not form, set test tube #1 aside and repeat the same procedure for the remaining four tubes.

3. Several of the reaction tubes may not have produced a precipitate If so, place these in a water bath at 45°C for another 10 minutes and record the time required for precipitation to occur. Note those tubes that do not produce a precipitate.

4. Record your results in Table 1 in the data and results section. Calculate the elapsed time (if any) and record this in Table 2 along with the structure of the appropriate alkyl halide.

Name _____ Sec. _____

DATA AND RESULTS

Table 1.

Name of the Alkyl Halide	Initial time (sec.)	Precipitation (sec.)	Elapsed time (sec.)
#1			
#2			
#3			
#4			
#5			

Table 2.

Structure of the Alkyl Halide	Elapsed time (sec.)
#1	
#2	
#3	
#4	
#5	

Name _____ Sec. _____

Place the five alkyl halides you tested in order of decreasing reactivity with silver nitrate in ethanol. Discuss in detail how this order is related to the structure of the individual substrates and the mechanism of the SN1 reaction.

QUESTIONS

1. Based on what you learned in this experiment, place the following alkyl halides in order of decreasing reactivity in SN2 reactions: vinyl chloride (chloroethene); isopropyl bromide; 2-chloro-2-methylpentane; and 2-bromo-2-methylpentane.

2. Draw the structure of (S)-2-bromo-2-phenylbutane and predict the product of its reaction with isopropyl alcohol. Show all stereochemistry.

3. It was observed that 1-bromo-1,1-diphenylmethane reacts many times faster than 1-bromo-1,1-dicyclohexylmethane. Draw structures and explain.

SECTION 5

SPECTROSCOPIC AND QUALITATIVE ANALYSES OF ORGANIC COMPOUNDS

Characterizing compounds is the most common procedure performed by the organic chemist. At this point in the course you have used various techniques to identify compounds prepared in the lab. This includes taking melting points and comparing them to literature values as well as using R_f values from thin-layer chromatograms. Various spectroscopic techniques are available for characterizing organic compounds. In lecture you will study several of these including nuclear magnetic resonance spectroscopy (NMR), infrared spectroscopy (IR), and mass spectroscopy (MS). In this section you will learn to use an FTIR instrument and to interpret the resulting IR spectra with respect to prominent functional groups that are present in various organic compounds. Along with melting points and TLC, you will use this technique routinely throughout the course as a criterion of purity of the various compounds you prepare.

In this section you will also learn some of the various wet methods available to the organic chemist for qualitatively identifying an organic compound or, at least, to determine the functional groups present on the molecule. There are various reactions available that are specific for aldehydes or esters or a variety of other compounds. Most of them involve color changes or precipitate formation that can be easily observed with microscale quantities of material. You will begin by doing some of these tests on known compounds, and later use them to identify various functional groups present on solid and liquid unknowns.

EXPERIMENT 30

QUALITATIVE ANALYSIS OF KNOWN ALKENES AND ALCOHOLS

INTRODUCTION

Note that many of the experiments in this lab manual involve reactants or products that contain a carbon carbon double bond or the alcohol functional group. Being able to identify the presence or absence of these and other functional groups is very important to the organic chemist. Although spectroscopic techniques have predominated in this area, older "wet methods" still have some value. In this experiment we will examine some of the reactions that can be used to identify the presence of an unsaturated carbon carbon bond. There are also methods that can be used to differentiate 1°, 2°, and 3° alcohols.

The presence of the alkene (or alkyne) functional group can be determined by mixing the unknown with a colored reagent that becomes colorless when it undergoes addition reactions with the substrate. The reactions are shown below and have already been covered in the lecture part of the course.

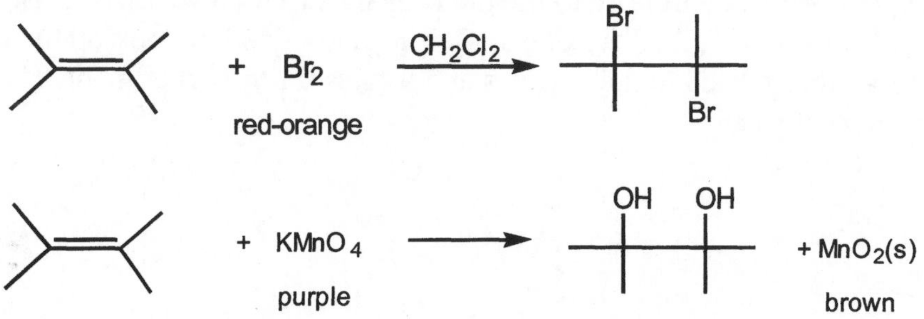

An excess of an alkene will decolorize a red-orange solution of bromine in methylene chloride. Dissipation of the purple color of an aqueous solution of potassium permanganate will occur when it oxidizes alkenes to vicinal diols. This is referred to as the Baeyer test, and over a period of time, a brown-black precipitate of manganese dioxide is often observed.

The presence of an alcohol can often be determined by its ability (or inability) to be oxidized to a ketone or carboxylic acid. A useful oxidizing agent is chromium trioxide dissolved in sulfuric acid. Often called the "Jones Reagent," the red-orange colored Cr^{+6} is reduced to Cr^{+3} often forming a precipitate of $Cr_2(SO_4)_3$ that can appear to be blue to green in color. The oxidation requires the loss of hydrogen from the carbinol carbon of the alcohol. Thus primary and secondary alcohols give a positive CrO_3 test observable by the loss of the orange color, darkening of the mixture, with the ultimate formation of the blue-green precipitate. Tertiary alcohols have NO carbinol hydrogens and thus do not react. A limitation to the test is that aldehydes are also oxidized to carboxylic acids.

$1°$, $2°$, $3°$ alcohols and aldehydes can be distinguished from one another using the "Lucas reagent." This is an aqueous solution of zinc chloride in concentrated hydrochloric acid. It generates a carbocation that combines with the chloride ion in an SN1 reaction. Because the resulting alkyl halide is not miscible with the aqueous part, the mixture separates into two layers that can be discerned by the appearance of cloudiness.

The rate of the reaction is related to carbocation stability. Tertiary alcohols produce an immediate cloudiness with the Lucas reagent, whereas secondary alcohols require 5 to 10 minutes for reaction to occur. Primary alcohols and aldehydes do not react. Of course, allylic and benzylic alcohols will react immediately also since they form resonance-stabilized carbocations.

In this experiment you will learn to use the reagents discussed to distinguish between alkanes, alkenes, and $1°$, $2°$, $3°$ alcohols. Before coming to lab, look up the hazards and safety precautions related to the reagents and solutes for the experiment. These can be listed on the next page.

Name _____ Sec. _____ Exp. # _____

Safety and Hazards

PROCEDURE

Indicate the particular test you are running in the Procedure column of the **procedure/observation** section. Then record your results in the observation column.

A. Br_2/CH_2Cl_2

Place 10 drops of each of the following in separate small test tubes: 1-hexene, hexane, 1-butanol, 2-butanol, and cyclohexene. Add five drops of the bromine/methylene chloride reagent to each tube one drop at a time and observe.

B. $KMnO_4$ (Baeyer Test)

Place 20 drops of each of the following in separate small test tubes: 1-hexene, hexane, 1-butanol, 2-butanol, and cyclohexene. Add five drops of the $KMnO_4$ reagent to each tube one drop at a time and observe.

C. CrO_3 / H_2SO_4 (Jones Reagent)

Place 20 drops of each of the following in separate small test tubes: 1-butanol, 2-butanol, tert-butyl alcohol, cyclohexane, benzaldehyde, and 1-hexene. Add five drops of the CrO_3 reagent to each tube one drop at a time and observe.

D. Lucas Reagent

Place 30 drops of the Lucas reagent into five test tubes. To the first tube add three drops of tert-butyl alcohol, to the second tube three drops of benzyl alcohol and observe. Add three drops of 2-butanol, 1-butanol, and benzaldehyde to the other three. Observe all five tubes every minute or so for the next 15 minutes.

Complete Table 1.

Name _____ Sec. _____ Exp. # _____

PROCEDURE	OBSERVATIONS

Name _____ Sec. _____ Exp. # _____

PROCEDURE	OBSERVATIONS

When known, write a **"+" or "-"** in the appropriate columns indicating whether the general solute gave a positive or negative result with the specified reagent.

Table 1. Results of Qualitative Organic Analysis of alkenes and alcohols

	Br$_2$	KMnO$_4$	CrO$_3$	Lucas
An alkane				
An alkene				
1° alcohol				
2° alcohol				
3° alcohol				
Benzyl alcohol				
An aldehyde				

EXPERIMENT 31

QUALITATIVE ANALYSIS OF UNKNOWN ALKENES AND ALCOHOLS

PROCEDURE

Your lab instructor will provide you with two liquid unknowns to identify using the qualitative organic analysis tests from the previous experiment. Each unknown may be an alkane, alkene, 1° alcohol, 2° alcohol, or 3° alcohol. Be sure to record the unknown letter for each one.

Br_2/CH_2Cl_2

Place 10 drops of each of the following in separate small test tubes: 1-hexene, hexane, and an unknown. Add five drops of the bromine/methylene chloride reagent to each tube one drop at a time and observe.

$KMnO_4$ (Baeyer Test)

Place 20 drops of each of the following in separate small test tubes: 1-hexene, hexane, and an unknown. Add five drops of the $KMnO_4$ reagent to each tube one drop at a time and observe.

CrO_3 / H_2SO_4 (Jones Reagent)

Place 20 drops of each of the following in separate small test tubes: 1-butanol, *tert*-butyl alcohol, and an unknown. Add five drops of the CrO_3 reagent to each tube one drop at a time and observe.

Lucas Reagent

Place 30 drops of the Lucas reagent into three test tubes. To the first tube add three drops of *tert*-butyl alcohol, to the second tube three drops of 1-butanol, and to the third an unknown. Observe your unknown every minute or so for the next 20minutes.

Record your observations and repeat the experiment for the second unknown. Your lab instructor will provide you with a list of possible unknowns. Using your observations, complete Table 1 at the end of this experiment. Using these results, as well as those tabulated in the previous experiment, identify your unknowns.

Name _____ Sec. _____ Exp. # _____

PROCEDURE	OBSERVATIONS

Name _____ Sec. _____ Exp. # _____

PROCEDURE	OBSERVATIONS

Name _____ Sec. _____ Exp. # _____

PROCEDURE	OBSERVATIONS

Table 1. Results of Qualitative Organic Analysis of Unknowns

	Br$_2$	KMnO$_4$	CrO$_3$	Lucas
hexane				
1-hexene				
1-butanol				
t-butyl alcohol				
Unknown ____				
Unknown ____				

Unknown Letter _____   Identity _____

Unknown Letter _____   Identity _____

QUESTIONS

1. How would you distinguish between cyclohexane and cyclohexene?

2. How would you distinguish between cyclohexanol and 1-hexanol?

EXPERIMENT 32

QUALITATIVE ANALYSIS OF KNOWN ALDEHYDES, KETONES, CARBOXYLIC ACIDS, AND ESTERS

INTRODUCTION

Many of the experiments in this lab manual involve reactants or products that contain an aldehyde, ketone, carboxylic acid, or ester functional group. Being able to identify the presence or absence of these functional groups is very important to the organic chemist. Although spectroscopic techniques have predominated in this area, older "wet methods" still have some value. In this experiment we will examine some of the reactions that can be used to identify the presence of these carbonyl-containing functional groups.

Aldehydes can be converted to carboxylic acids using a mild oxidizing agent that will not react with alcohols. The most common one is the Tollens' reagent, an ammoniacal silver ion solution. The Ag^+ is reduced to silver metal as it oxidizes the aldehyde. This metal appears as a "silver mirror" on the side of the test tube thus providing a positive test for the presence of an aldehyde. Most ketones do not react with the Tollens' reagent. The unbalanced equation for this oxidation is shown below.

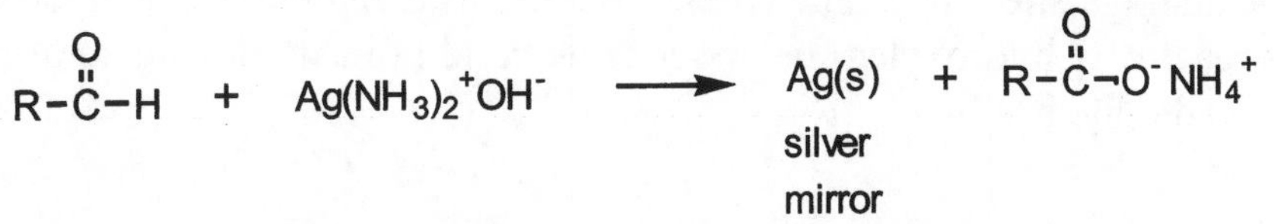

As you learned in lecture, these aldehydes and ketones react with 2,4-dinitrophenylhyrazine to form a highly colored 2,4-dinitrophenylhydrazone solid. The formation of this yellow or orange precipitate indicates the presence of an aldehyde or ketone. It is often called a "2,4-DNP derivative." Carboxylic acids and esters do not react.

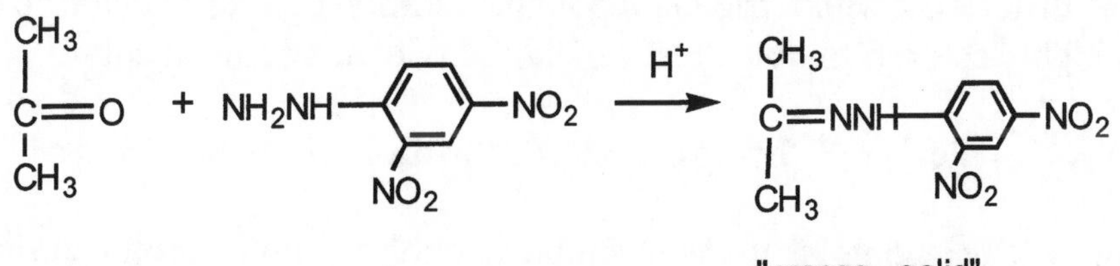

The melting point of a DNP-derivative can often be used to identify a specific aldehyde or ketone by comparison to literature values. For example, the DNP-derivative of acetone melts at 125°C.

265

Aldehydes can sometimes be distinguished using the Schiff's reagent, a colorless derivative of a Fuchsin dye. When mixed with an aldehyde, a violet color slowly appears as it reacts with the dye. Unhindered ketones can also react, although normally at a slower rate.

As you saw in one of your enolate chemistry lectures, a mixture of iodine and sodium hydroxide will react with methyl ketones producing a carboxylic acid and CHI_3, a yellow precipitate called iodoform. Other ketones do not react.

$$R-\overset{\overset{\displaystyle O}{\|}}{C}-CH_3 \xrightarrow[\text{NaOH}]{I_2} R-\overset{\overset{\displaystyle O}{\|}}{C}-Cl_3 \xrightarrow{\text{NaOH}} R-\overset{\overset{\displaystyle O}{\|}}{C}-O^- \ Na^+ + CHI_3$$

Acetaldehyde (CH_3CHO), a "methyl aldehyde" will also react. 2-methylalkanols will also give a positive iodoform test since many of them are oxidized by iodine to methyl ketones.

$$R-\overset{\overset{\displaystyle OH}{|}}{CH}\cdot CH_3 \xrightarrow[\text{NaOH}]{I_2} R-\overset{\overset{\displaystyle O}{\|}}{C}-CH_3$$

Esters can be distinguished from aldehydes and ketones using the "ferric hydroxamate test." Esters react with hydroxylamine under basic conditions producing an organic hydroxamic acid ("hydrox").

$$R-\overset{\overset{\displaystyle O}{\|}}{C}-OR + NH_2OH \longrightarrow R-\overset{\overset{\displaystyle O}{\|}}{C}-NHOH$$
$$\text{"hydrox"}$$

The "hydrox" then reacts with ferric ion upon the addition of a ferric chloride solution producing a highly colored complex that can be red, violet, or blue in color.

$$Fe^{+3} + 3 \text{ "hydrox"} \rightarrow Fe(\text{"hydrox"})_3$$

The ferric ion will also complex to the phenolic alcohol functional group producing a highly colored complex. This is often used to test for the presence of a phenolic hydroxyl group in an unknown.

Carboxylic acids do not react with the aforementioned reagents. Its presence is usually determined from solubility experiments. Unknown carboxylic acids not soluble in water will dissolve in a saturated $NaHCO_3$ solution via the formation of its sodium salt. See experiment 2 for details.

Safety and Hazards

PROCEDURE

Indicate the particular test you are running in the procedure column of the procedure/observation section. Then record your results in the observation column.

A. Tollens' Test

The Tollens' reagent needs to be prepared just before it is used. Three milliliters of the reagent can be prepared by adding two drops of 5% NaOH to 1 ml of a 5% $AgNO_3$ solution. This produces a precipitate of Ag_2O. Then add 3M NH_3 to the tube until the precipitate just dissolves. Add 10 drops of the Tollens' reagent to five separate test tubes. Then add two drops of each of the following solutes: benzaldehyde, 2-butanol, 1-hexene, acetaldehyde, and acetophenone. Wait 5 to 10 minutes for the silver mirror to form. Place those tubes that do not react in a hot water bath for 10 minutes and observe.

B. 2,4-Dinitrophenylhydrazone Derivative

Add 10 drops of ethanol to five separate test tubes. Then add 10 drops of benzaldehyde, 2-butanol, methyl salicylate, and acetophenone individually to four of the tubes. In the fifth tube dissolve a small amount (about the size of a grain of rice on a microspatula) of benzoic acid. Add 10 drops of the 2,4-dinitrophenylhydrazine reagent. The immediate formation of a yellow or orange precipitate indicates the presence of an aldehyde or ketone.

C. Schiff Test

Place 20 drops of the Schiff reagent in five test tubes. Then add five drops of benzaldehyde, 2-butanone, 2-butanol, acetaldehyde, and methyl salicylate separately to each one. Set aside for about an hour. Shake and observe every 5 minutes. Violet or reddish-violet color suggests the presence of an aldehyde.

D. Iodoform Test

Add five drops of acetophenone, benzaldehyde, 2-butanol, cyclohexanone, and 2-butanone to five separate test tubes. Then to each add 10 drops of water, shake, and add 10 drops of 5% NaOH. Then add 20 drops of the Iodine reagent dropwise with shaking. Formation of a yellow precipitate is a positive test.

E. Ferric Hydroxamate test

Add two drops of acetophenone, 2-butanol, and methyl salicylate to three separate test tubes. Add 15 drops of a 0.5M alcoholic soluton of hydroxylamine and five drops of a 10% NaOH solution. Boil the alcoholic solution on a steam cone, cool, and neutralize the NaOH with 30 drops of a 1M HCl solution. Add ethanol dropwise until the mixture is clear. Then add the $FeCl_3$ reagent dropwise and observe. A red to violet color is considered a positive test.

Name _____ Sec. _____ Exp. # _____

PROCEDURE	OBSERVATIONS

Name _____ Sec. _____ Exp. # _____

PROCEDURE	OBSERVATIONS

Name _____ Sec. _____ Exp. # _____

When known, write a **"+" or "-"** in the appropriate columns indicating whether the general solute gave a positive or negative result with the specified reagent.

Table 1. Results of Qualitative Organic Analysis Knowns.

	Tollens	2,4-DNP	Schiff	Iodoform	"hydrox"
2-butanol					
acetophenone					
benzaldehyde					
2-butanone					
Methyl salicylate					
Benzoic acid					
1-hexene					
Acetaldehyde					
cyclohexanone					

EXPERIMENT 33

QUALITATIVE ANALYSIS OF UNKNOWN ALDEHYDES, KETONES, CARBOXYLIC ACIDS, ESTERS, AND PHENOLS

PROCEDURE

Your lab instructor will provide you with one liquid and one solid unknown to identify using the qualitative organic analysis tests from the **previous experiment**. Each unknown may be an aldehyde, ketone, carboxylic acid, ester, or a phenol. Be sure to record the unknown letter for each one. You need to use about 20 mg of the solid unknown. This can be done by weighing the first time you use it, then comparing it to the size of a grain of rice, or pea, and so on, and just estimating the amount for later tests.

For each case set up three test tubes for the determination: one that you know will give a positive test, one that will be negative, and your unknown. This procedure is used so that you know the reagents have not decomposed or have become contaminated. You may begin by running the Tollens', 2,4-DNP, Schiff's, iodoform, and ferric hydroxamate tests. Before testing an ester, you need to add just a ferric chloride solution to your test tube. Many phenols will give a highly colored complex with Fe^{+3}. If this test is positive, then you will not be able to do the hydroxamate test. If all tests are negative, then you need to use experiment 2 to do the solubility tests for carboxylic acids.

Record your observations and repeat the experiment for the second unknown. Your lab instructor will provide you with a list of possible unknowns. Using your observations, complete Table 1 at the end of this experiment. Using these results, as well as those tabulated in the previous experiment, identify your unknowns.

Name _____ Sec. _____ Exp. # _____

PROCEDURE	OBSERVATIONS

PROCEDURE	OBSERVATIONS

Name _____ Sec. _____ Exp. # _____

PROCEDURE	OBSERVATIONS

Name _____ Sec. _____ Exp. # _____

Table 1. Results of Qualitative Organic Analysis Unknowns

	Tollens	2,4-DNP	Schiff	Iodoform	"hydrox"
Liquid unknown __					
Solid unknown ___					

Liquid unknown letter _____ Identity _____

Solid unknown letter _____ Identity _____

QUESTIONS

1. How would you distinguish between pentanal and 2-pentanone?

2. How would you distinguish between pentanoic acid and n-pentylpentanoate?

EXPERIMENT 34

QUALITATIVE ANALYSIS OF A GENERAL UNKNOWN

PROCEDURE

Your lab instructor will provide you with one liquid and one solid unknown to identify using the qualitative organic analysis tests from the previous experiments. Each unknown may be an Alkene, alcohol, aldehyde, ketone, carboxylic acid, ester, or a phenol. Be sure to record the unknown letter for each one. You need to use about 20 mg of the solid unknown. This can be done by weighing the first time you use it, then comparing it to the size of a grain of rice, or pea, and so on, and just estimating the amount for later tests.

For each case set up three test tubes for the determination: one that you know will give a positive test, one that will be negative, and your unknown. This procedure is used so that you know the reagents have not decomposed or have become contaminated. You may begin by running the typical tests you did with alkenes and alcohols followed by the Tollens', 2,4-DNP, Schiff's, iodoform, and ferric hydroxamate tests. Before testing an ester, you need to add just a ferric chloride solution to your test tube. Many phenols will give a highly colored complex with Fe^{+3}. If this test is positive, then you will not be able to do the hydroxamate test. If all tests are negative, then you need to use experiment 2 to do the solubility tests for carboxylic acids.

Record your observations and repeat the experiment for the second unknown. Your lab instructor will provide you with a list of possible unknowns. Using your observations, complete Table 1 and 2 at the end of this experiment. Identify your unknowns.

Name _____ Sec. _____ Exp. # ____

PROCEDURE	OBSERVATIONS

Name _____ Sec. _____ Exp. # _____

PROCEDURE	OBSERVATIONS

Name _____ Sec. _____ Exp. # _____

PROCEDURE	OBSERVATIONS

Name _____ Sec. _____ Exp. # _____

Table 1. Results of Qualitative Organic Analysis of Unknowns

	Br_2	$KMnO_4$	CrO_3	Lucas
hexane				
1-hexene				
1-butanol				
t-butyl alcohol				
Unknown ____				
Unknown ____				

Table 2. Results of Qualitative Organic Analysis of Unknowns

	Tollens	2,4-DNP	Schiff	Iodoform	"hydrox"
Liquid unknown __					
Solid unknown __					

Liquid unknown letter _____ Identity _____

Solid unknown letter _____ Identity _____

EXPERIMENT 35

NUCLEAR MAGNETIC RESONANCE SPECTROSCOPY

INTRODUCTION

The simple definition of spectroscopy is the study of the interaction of electromagnetic radiation with matter. Nuclear magnetic resonance (NMR) spectroscopy is the use of this phenomenon to study chemical and biological properties of matter. It is an indispensable tool routinely used in organic chemistry to determine the structure of organic molecules. Organic chemists widely use NMR spectroscopy to study simple structures utilizing a one dimensional technique; however, two-dimensional techniques are used to analyze complicated structures. The flexibility of NMR makes it invasive in the sciences, especially in the biological and chemical field. In this laboratory experiment, it is essential to have a basic background of NMR spectroscopy.

Scientists found that certain nuclei such as ^{1}H. ^{13}C, ^{18}O, ^{31}P, ^{15}N, and ^{19}F behave like tiny magnets. When these magnets are exposed to a magnetic field, they assume two energy states. One energy state is aligned to the magnetic field (lower in energy), and the other energy state is opposed to the magnetic field (higher in energy). The energy difference in these two states is quantized. The energy is absorbed or emitted at certain radio frequencies. These radio frequencies depend on the magnetic field and the type of the nucleus used in the analysis. Probably the most important reason is the fact that slightest difference in the electronegativity and bonding state of surrounding atoms cause small variations (parts per million) in the magnetic filed felt by each nucleus. The term, chemical shift, is used to denote a small variation, and plotted versus signal intensity to obtain an NMR spectrum. The interpretation of these signals and other special features of the NMR spectrum is a special skill that can be acquired after solving NMR spectral problems.

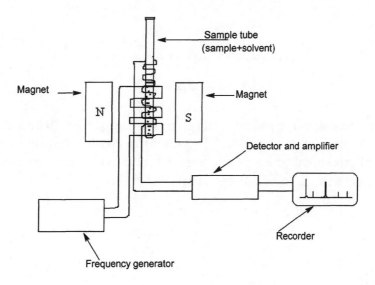

PRACTICAL ASPECTS

In general, NMR spectrometers constitute a magnet, a detector, a recorder, and source of radio frequency energy. The sample prepared in a proper solvent is placed in the cavity located between two magnets. Then the sample is scanned by two different methods. In the beginning of the development of NMR techniques, the frequency of the instrument was kept constant while the sample was scanned. In contrast, in advanced methods, the frequency is changed while the field is kept constant. Also, the newer instruments can perform multiple scans to reduce noise and enhance the sensitivity to provide better plots.

In the beginning of the advancement of the spectroscopy field, the instruments were 60 MHz (megahertz). Now, more powerful instruments are available for research and educational purposes: 90, 100, 220, 300, 400, or 500 MHz. To acquire better resonance, the intense magnetic field is required. As a result, the frequency is increased. This higher frequency is useful to get improved resolution of peaks and simplified spectra of complex molecules

BASIC NMR PLOT

A typical NMR spectrum is a graph of absorption (on the y-axis) as a function of the applied magnetic field (on the x-axis). The x axis values can be presented in frequency as MHz or in δ (ppm).

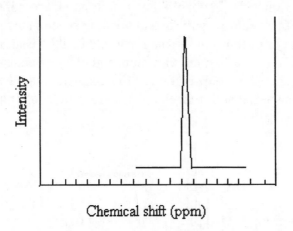

Chemical shift (ppm)

The x-axis represents chemical shift (ppm) starting at right-hand side as 0.0ppm, and it extends to 14 ppm on the left side. The y-axis represents intensity of the signal. The integration of this signal (the area under the peak) provides the spectroscopist with the number of protons in the molecule producing the signal.

Units Review

In NMR spectroscopy, scientists use a set of units to describe magnetic field, temperature, energy, frequency, and degrees of angles. These units are compiled for better understanding of the basics of NMR spectroscopy.

1. The frequency of electromagnetic radiation may be reported in cycles per second (radians per second). Frequency in cycles per second (Hz) have units of inverse seconds (s^{-1}). It is noted as v or f.
2. The conversion between Hz and rad/s is trivial. There are 2π radians in a cycle. Therefore, $1 \text{ Hz} = 1 \text{ s}^{-1} = 2\pi \text{ rad/s}$
3. Unit of time is seconds (s).
4. Magnetic filed strength (B) is measured in Tesla (T).
5. The unit of energy (E) is the Joule (J). Note: Chemists represent the relative energy of a particle using an energy level diagram in NMR.
6. Generally, angles are reported in degrees ($^{\circ}$) and in radians (rad).
7. The absolute temperature scale Kelvin (K) is used in NMR.
8. Power is the energy consumed per time, and it is expressed in Watts (W).

CHEMICAL SHIFTS

1. Definition
2. Downfield and upfield
3. Shielded and unshielded protons
4. Assignment
5. Table of typical values of chemical shift

1. Definition
The chemical shifts are the variations in the positions of NMR absorptions resulting from electronic shielding and deshielding.

2. Downfield and Upfield

Before an NMR is performed, the instrument is set to "0 ppm" with an internal standard. Usually trimethylsilane (TMS) is used as the standard because most protons in organic compounds absorb at a higher frequency than TMS. Conventionally, protons that absorb at the higher frequency (to the left in the spectrum) are said to be "downfield," and those that absorb at lower frequencies (to the right in the spectrum) closer to TMS are said to be "upfield."

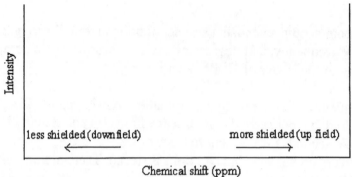

less shielded (downfield)

more shielded (up field)

Chemical shift (ppm)

3.   Shielded and unshielded protons

The resonance frequency of a proton is determined by its electronic environment. The electrons of the hydrogen generates its own small magnetic field that "shields" the nucleus of the proton from the applied magnetic field of the instrument. The more shielded a proton is the further upfield the absorption peak will appear. Electron-withdrawing groups near the proton of interest will "deshield" the hydrogen, shifting the resonance frequency downfield. This property plays a major role in the 3NMR spectrum and yields spectrum at different chemical shifts. It is necessary to know how to find shielding and deshielding of protons because it implies the electronic structure of the molecule. The major factor is electronegativity of an atom to which the hydrogen is attached to or resides close to it.

H

H

Less shielded
proton

More shielded
protons

H—C—O

absorb at a
lower field
(left hand side)

absorb at a
higher field
(right hand side)

H

The more electronegative a substituent, and the closer it is to the proton of interest, the greater the deshielding effect is observed in NMR plot.

4.   Assignment of protons

The more shielded proton appears on the right side of the plot (upfield)
The less shielded proton gives a signal on the left side of the plot (downfield). An approximate specific organic functional group proton assignment is given as follows:

288

Proton group	Chemical shift range (ppm)	Proton group	Chemical shift range (ppm)
Tetramethyl silane	0.0	Br – CH	2.5-3.8
– CH_3 (saturated 1°)	0.7-1.3	Cl –CH	2.8-3.8
R_2NH	0.8-3.0	RO –CH (ether)	3.0-4.0
–CH_2R (saturated 2°)	1.2-1.4	HO – CH	3.0-4.0
–CH (saturated 3°)	1.4-1.7	R –COO–CH	3.5-4.5
$R_2C=CR–CH$	1.5-2.0	Ph O – CH	3.5-5.5
NC–CH	1.5-3.0	F –CH	4.0 - 4.5
R – OH	1.5-5.5	NO_2 –CH	4.0-4.5
$C \equiv C – H$	1.6- ~2.4	Ph – OH	4.0-8.0
I –CH	1.8-2.8	$R_2C = CH_2$	4.5-5.5
RCO – CH	1.8-2.8	RCH = CHR	5.0-6.5
$R_2 N–CH$	1.8-3.0	RCONH(amide)	5.5-8.5
RCOOH	10.0-12.5	Aromatics	6.0-8.5
Ar – CH	2.0-2.5	RCHO	9.5-10.0
RS – CH (sulfide)	2.0-3.0		

A.   Splitting Patterns and Coupling Constants

Splitting patterns and coupling constants in the NMR spectrum help to determine group locations in a molecule's carbon skeleton. More specifically, it helps to find which groups are located next to each other. If one or more equivalent protons couple to one adjacent proton, then the coupling hydrogen appears as a doublet in the intensity line separated by the coupling constant (J). "J" is measured in Hertz (Hz). The triplet of lines is observed if the coupling proton or protons couple equally to two protons three chemical bonds away. The relative intensities of 1:2:1 is recorded on the NMR plot. In general, chemically equivalent protons give a pattern of lines containing one more line than the number of protons being coupled to, and the intensities of the peaks follow the binomial expansion. It can be said that it splitting can be presented by n +1.

However, all coupling does not follow n +1 rule. In these special cases the multiplets are very complex and difficult to interpret. Therefore, these samples are run on the higher frequency instrument that convert these complex pattern into "first order," n +1 multiplets. The size of J-values provides important structural information of the molecule.

To learn the basic mechanics of obtaining and interpreting an NMR spectrum
If time permits each pair can do one unknown.

B.   Integration

Integral values are provided with spectrum and that helps to determine the number of hydrogens present on the carbon. The heights of the steps are proportional to the number of protons in that group, and it is helpful to find exact number of hydrogens.

PROCEDURE

Please record the number of the assigned unknown in the lab notebook. According to your instructor's instructions, dissolve sample in a given solvent. Go to the NMR room with your handout. In the presence of your instructor, acquire NMR plot and make extra one copy of it for your record. Attach this NMR plot to your lab report.

SAMPLE PREP

A routine sample for a proton NMR on a 400 M-Hz instrument consists of one to two drops of the sample or 10 mg of the compound in about 0.7-1mL of solvent in a 5mm NMR tube.

Solvent

Generally, dimethyl sulfoxide, deuterated chloroform, acetone, tetramethylsilane are used to acquire NMR plots. There are some standard properties observed in choosing solvent. It should not contain protons, should be inert, low boiling point, deuterated, and inexpensive.

Common solvents are TMS (tetramethylsilane) and deuterated chloroform ($CDCl_3$). Deuterated chloroform ($CDCl_3$): widely used, peak at $\delta = 7.26$ is observed due to CHCl3 impurity.
TMS (tetramethylsilane) $(CH_3)_4$ Si: the most useful reference compound for its properties: Chemically inert to most reagents, symmetrical, volatile (b.p. $26.5^{\circ}C$), soluble in most organic molecules, gives a single, intense, sharp absorption peak. Its protons are more shielded than almost all organic protons.

The NMR tube must be cleaned after each use immediately. The caps can not be reused since it is a source of contamination. Always use a new cap for each sample prepared for analysis. The cap is necessary on the NMR tube to prevent evaporation of the solvent.

The NMR tube must be wiped with a lint-free tissue before inserting it into the probe. The directions to operate computer software are provided in the NMR room or by your instructor.

NMR INTERPRETATION

1. Use the molecular formula to determine the units of unsaturation
2. Observe and measure chemical shifts
3. Assign peaks: check relative peak areas, intensities of the signals, splitting of the signals
4. write down fragment parts of the molecule
5. assemble molecule structure

Consider the following example and apply five steps to obtain molecular structure.

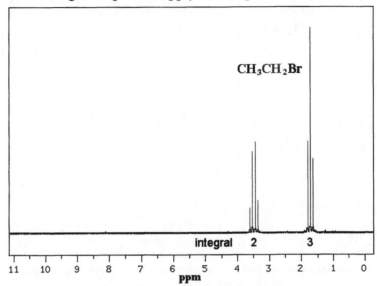

Determination of units of unsaturation

Use the following general formula to calculate degrees of unsaturation:

$$\text{Degrees of unsaturation} = [(\text{number of carbon atoms} \times 2) + 2 - (\text{number of hydrogens})] \div 2$$

- Oxygen- ignore
- Halides (F, Cl, Br)- count as a hydrogen
- Nitrogen - count as half of a carbon
- Double bond count as 1 degree of unsaturation
- Aromatic ring 1 degree of unsaturation
- Triple bond count as two degrees of unsaturation

Observe chemical shifts and write down molecular fragments

HINT: It is useful to write down chemical shifts in table format at the beginning.

Chemical shift [ppm]	Number of peaks	Integration	Fragments of molecule
3.5 (less shielded)	4 (next carbon has three hydrogens)	2 (carbon has two hydrogens)	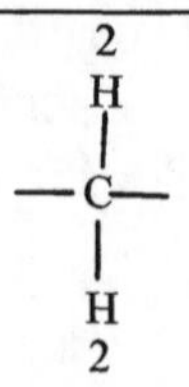 2 H | —C— | H 2
1.8 (more shielded)	3 (next carbon has two hydrogens)	3 (carbon has three hydrogens)	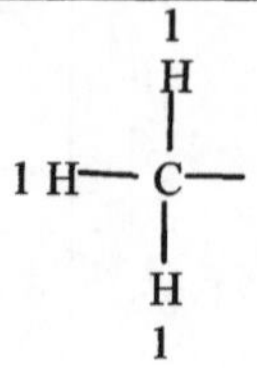 1 H | 1 H—C— | H 1

5. Assemble the molecular structure

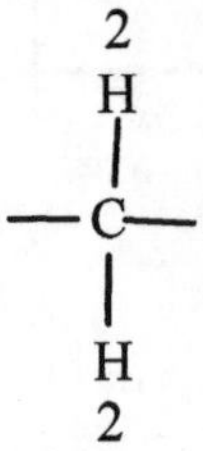

```
      2
      H
      |
    — C — Br   must present beacuse we have 4 peaks in NMR plot due to
      |              3 hydrogens in neighborhood
      H
      2
```

Final structure of the molecule can be written as shown below:

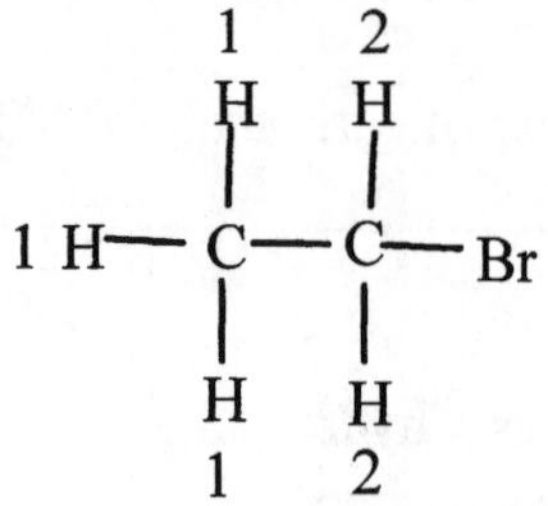

```
       1     2
       H     H
       |     |
  1 H— C ——— C —Br
       |     |
       H     H
       1     2
```

EXPERIMENT 36

BOROHYDRIDE REDUCTION OF
2- METHYLCYCLOHEXANONE AND MOE
ANALYSIS OF THE PRODUCT

INTRODUCTION

In general, compounds containing aldehyde or ketone functional groups play a significant role in organic chemistry synthesis. These carbonyl compounds are used in the synthesis of esters, acid halides, carboxylic acids, alcohols and so on. One of the useful reactions in the synthesis of alcohols is the reduction of ketones and aldehydes. More specifically, aldehydes and ketones are easily reduced to the corresponding primary and secondary alcohols, respectively.

This synthesis can be accomplished by using a variety of reducing agents. For laboratory applications, among the most useful reagents providing excellent yields are the complex metal hydrides such as sodium borohydride ($NaBH_4$) and lithium aluminum hydride ($LiAlH_4$).

sodium borohydride	lithium aluminum hydride

Careful examination of the structures of $NaBH_4$ and $LiAlH_4$ shows that hydrogen atoms are covalently bonded to boron and aluminum atoms with negative charges and makes complex metal hydrides better nucleophiles and reduces their basicity.

Of these two reagents, lithium aluminum hydride is much stronger, reactive and difficult to handle in the laboratory. It reduces ketones and aldehydes; and other functional groups such as acids and esters. On the other hand, sodium borohydride ($NaBH_4$) offers certain advantages in the laboratory because it is moderately stable in aqueous and alcoholic solutions as long as the solution is basic (high pH). As a result, sodium borohydride is considered a convenient and selective reducing agent. It can reduce a ketone or an aldehyde using a wide variety of solvents, including alcohols, ethers, and water in the presence of an acid and ester functional groups. This is not possible in the case of lithium aluminum hydride.

In this experiment, an alcohol is synthesized using sodium borohydride to reduce a secondary ketone. In detail, the reduction process of 2-methylcyclohexanone is carried out in a protic solvent, methanol, at low temperature, preferably in the ice bath. The addition of a strong base is required to decompose the borate ester. Moreover, extraction and evaporation processes are used to eliminate the remaining solvent, leaving pure 2-methylcyclohexanol behind.

In the reduction of 2-methylcyclohexanone, the (cis- and trans-) isomers of 2-methylcyclohexanol are formed. These isomers are formed as follows:

2-methylcyclohexanone   Cis   Trans

2-methylcyclohexanol

This is simply a hydride reduction of a ketone through a nucleophilic addition. In this reaction hydride ion (H: $^-$) serves as the nucleophile. Attack by hydride gives an alkoxide that becomes protonated to an alcohol in the presence of a protic solvent. In this particular synthesis of 2-methylcyclohexanol, two isomers are formed (*cis/trans*) due to the chirality of the starting material, 2-methylcyclohexanone. Attack of the hydride from the back face, *trans* to the methyl group will produce the *cis* product alcohol. On the other hand, if the hydride attacks from the front face, the *trans* product alcohol will form.

Note that the reactant has gained hydride ion from the reducing agent, NaBH$_4$ and a proton from the protic solvent, methanol. In addition, both the *cis* and *trans* isomers of the 2-methylcylochexanol exist in four chair conformations. It is interesting to know that the formation of *cis*- and *trans*- isomers in this reduction process does not take place as a 50:50 mixture exactly. Therefore, it is necessary to acquire the NMR of the product and find the integration values that give the actual percentage of products in this reaction.

OBJECTIVES

1. The purpose of this experiment is to synthesize 2-methylcyclohexanol and determine the percent yield of product. In addition, the percent yield in this reaction can be determined by using nuclear magnetic resonance (NMR) spectroscopy.
2. Furthermore, using a molecular mechanics program, molecular operational environment (MOE), the steric energies of four chair conformations can be calculated and the most stable chair conformation predicted.
3. To confirm purity of the product and to determine whether the reduction of 2-methylcyclohenone has been completed, infrared (IR) spectrum of the product is useful in this experiment.

General mechanism is for understanding mole consumption and calculation of theoretical and percent yield in this reaction.

NOTE: for each mole of NaBH$_4$ 4 moles of hydride provided reducing 4 moles of ketone overall. In calculation of percent yield, this fact must be considered. In conclusion, 2-methylcyclohexanone is a limiting reagent.

1.

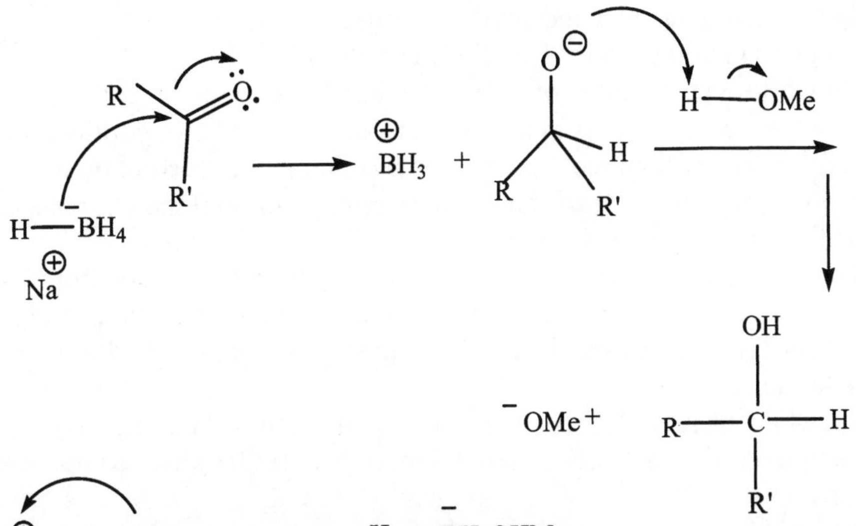

2.

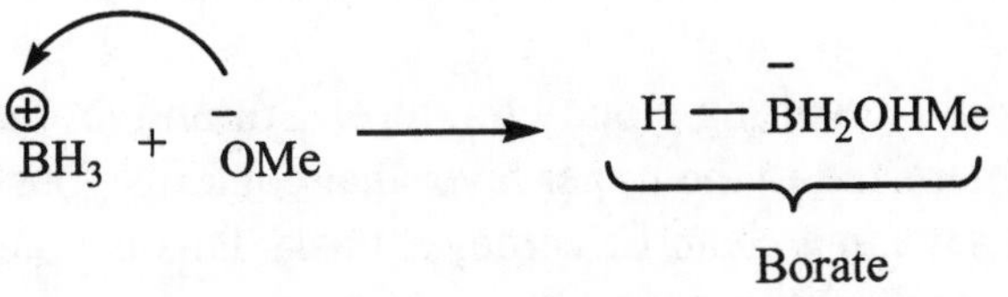

Borate

The acid-base reaction of methoxide and BH$_3$ yields a borate complex that has three more remaining B-H bonds. Further, each of these reduces another molecule of the stanrting material, ketone. Therefore, 1 mole of sodium borohydride reduces 4 moles of ketone overall.

3.

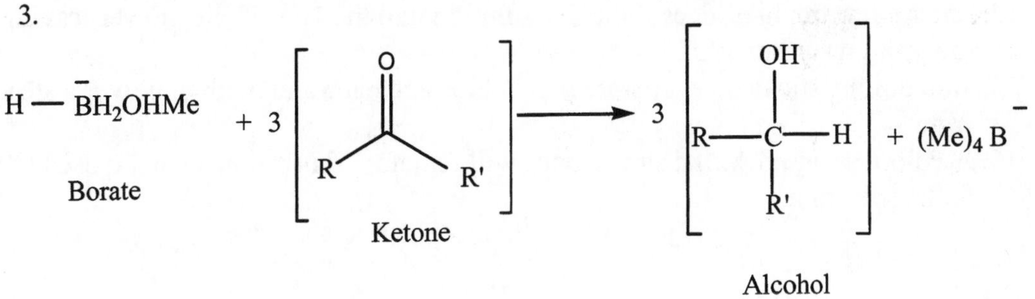

Borate   Ketone   Alcohol

PROCEDURE

1. Tare a 15×125mm test tube in 50 ml Beaker
2. Weigh 600mg of 2-methylcyclohexanone in this reaction tube and note down exact amount.
3. Add 2.5 ml of methanol and swirl few times.
4. Cool this solution in an ice bath for 5 minutes.
5. Meanwhile, weigh 100mg of sodium borohydride.
6. While the reaction tube is in the ice bath, add sodium borohydride in the flask carefully. You will notice the vigorous reaction in the reaction tube. Leave this reaction tube in the ice bath for 15 to 20 minutes or until the vigorous reaction has ceased. Stir the reaction using a stirring rod.
7. When the vigorous reaction is ceased, remove the reaction tube from the ice bath and leave at room temperature for 15 minutes. The reaction should appear to be finished. If it is not, leave the reaction mixture at room temperature for another five minutes.
8. Meanwhile, measure 2.5 ml of 3M NaOH and 2 ml water separately.
9. After the reaction is finished, add 2.5 ml of NaOH (3M) to decompose borate ester.
10. Later, add 2 ml of water to the resulting cloudy mixture. Stir and let the two layers separate. You are interested in the upper layer that contains your product.
11. Remove the top layer and save in a clean Erlenmeyer flask. This is your product. Add 2 ml of methylene chloride and swirl.
12. Add a small amount of sodium sulfate to the Erlenmeyer and swirl for a few minutes.
13. Weigh a 10-ml Erlenmeyer flask and record the weight in your data table.
14. After a few minutes, decant the dry solution into the 10-ml Erlenmeyer leaving the sodium sulfate behind.
15. Add boiling stick and evaporate the dichloromethane and methanol over a steam cone.
16. A colorless liquid with a strong odor will remain. This product can be used to run an IR spectrum.

IR Spectrum:
1. Obtain the IR spectrum of the product and label peaks
2. Locate major signals and check if reduction has been completed.

Computational Chemistry: (Model building)

Use the computational chemistry (MOE) handout to finish this part

Build a model of *cis* and *trans* 2-methylcyclohexanol using MOE
1. Print out your molecular models with steric energies and attach to your lab report for grading
2. Find out which isomer is more stable
3. Does this pattern fit your predictions?

NMR Spectrum

1. From the NMR plot, determine the integrations and chemical shifts for *cis*- and *trans*- products.
2. Calculate the trans/cis product ratio

Name _____ Sec. _____ Exp # _____

REACTANT TABLE

Name of compound	m.p./ b.p.	FW	Mass (g)	Vol. (ml)	D (g/ml)	Moles

SAFETY/HAZARDS

Name of compound	**Safety/ hazard**

Name _____ Sec. _____ Exp # _____

PROCEDURE	OBSERVATIONS

Name _____ Sec. _____ Exp # _____

PROCEDURE	OBSERVATIONS

Name ______________________________ Sec. __________ Exp # ________

DATA and CALCULATIONS

Table 1

Weight of Erlenmeyer flask and 2-methylcyclohexanol (g)	
Weight of empty Erlenmeyer fflask	
Weight of 2-methylcyclohexanol (g)	
Theoretical yield of 2-methylcyclohexanol (g)	
Percent yield	

Write the equation for this reduction process and show all calculations.

Equation:

Theoretical yield:

Percent yield:

Name _____ Sec. _____ Exp # _____

IR Spectrum:

1.

Major peaks

2. Explain in few lines with valid reasons whether a reduction reaction took place or not.

NMR Spectrum:

1. Integration values

Cis isomer _____   Trans isomer _____

2. Show calculation of trans/cis ratio

COMPUTATIONAL CHEMISTRY

Draw and label the structures of the two different products in the four possible chair conformations. Indicate the steric energy of each as determined using MOE program.

Predicted stable chair conformer _____

1. Steric energy _____	**2.** Steric energy _____
3. Steric energy _____	**4.** Steric energy _____

Name _____   Sec. _____   Exp # _____

List the most stable structure based on your computational data analysis.

The most stable structure   _____

Explain your reason briefly for the stability of this isomer.

EXPERIMENT 37

THE REDUCTION OF 2-METHYLCYCLO-HEXANONE AND THE STEREOCHEMICAL ANALYSIS OF THE PRODUCT

In this experiment, a secondary alcohol is synthesized using sodium borohydride to reduce a ketone. In detail, the reduction process of 2- methylcyclohexanone is carried out in a protic solvent, methanol, at lower temperature, preferably in the ice bath. The addition of a strong base is required to decompose the borate ester. Extraction and evaporation processes are used to eliminate the remaining solvent, leaving 2-methylcyclohexanol behind.

When hydride (H:) attacks an sp^2-hybridized, trigonal planar carbonyl group, it can approach from either the front (wedged) or back (dashed).

If the product is not chiral, only one product will form. However, in this experiment you will be reducing 2-methylcyclohexanone in which carbon #2 is chiral. Your starting material will be a racemic mixture of 2-methylcyclohexanone. Note that the product has two stereocenters, one on carbon #1 and the other on carbon #2. Therefore, there are four possible products: (R,R), (S,S), (1R,2S), and (1S,2R). In preparation for this experiment before you come to lab, complete the diagram on the next page. Be sure to label the correct configurations and geometrical isomers.

There are two parts to this experiment. First, you will do the synthesis in the laboratory, determine a yield and percent yield, and characterize the product by taking an IR. The reaction itself will produce a mixture of the *cis* and *trans* isomers. In the second part, you will examine an NMR of the product and determine the ratio of *cis/trans*.

In addition, both the *cis* and *trans* isomers of the 2-methylcylochexanol exist in four chair conformations. It is interesting to know that the formation of *cis*- and *trans*-isomers in this reduction process does not take place as a 50:50 mixture exactly. Therefore, it is necessary to acquire the NMR of the product and find the integration values that give the actual percentage of products in this reaction.

How are the two *cis* isomers related to one another? _____

Name _____ **Sec.** _____

Reactant Table

Name of compound	m.p./b.p.	F.W.	Mass (g)	Vol. (ml)	D (g/ml)	moles

Safety/Hazards

PROCEDURE	OBSERVATIONS

Name _____ Sec. _____ Exp. # _____

PROCEDURE	OBSERVATIONS

PROCEDURE

17. Place a 15- x 125-ml test tube from your microscale kit in a 50-mL beaker and tare them.
18. Weigh 600mg of 2-methylcyclohexanone in the reaction tube and note the exact amount.
19. Add 2.5 ml of methanol and swirl
20. Cool this solution in an ice bath for 5 minutes.
21. Meanwhile, weigh 100 mg of sodium borohydride.
22. While the reaction tube is in the ice bath, add sodium borohydride carefully. You will notice the vigorous reaction. Leave the mixture in the ice bath for 15 to 20 minutes or until the reaction has ceased. Stir with a stirring rod.
23. Remove the reaction tube from the ice bath and leave at room temperature for 15 minutes. The reaction should appear to be finished.
24. Add 2.5 ml of NaOH (3M) to decompose the borate ester.
25. Add 2 ml of water to the resulting cloudy mixture. Stir and let the two layers separate.You are interested in the upper layer that contains your product.
26. Remove the top layer with a Pasteur pipette and transfer it to a clean 10-ml Erlenmeyer flask. This is your product.
27. Add 2 ml of methylene chloride and dry with anhydrous sodium sulfate.
28. Meanwhile, weigh another 10-ml Erlenmeyer flask and record the weight.
29. After a few minutes, decant the dry solution from step # 11 into the weighed Erlenmeyer flask.
30. Add a boiling stick and evaporate the dichloromethane and methanol gently over a steam cone.
31. A colorless liquid with a strong odor will remain. Determine the yield and do an IR of the product.

Name _____ Sec. _____

DATA and Results of the Synthesis

Weight of product: _____

Calculation of percent yield:

Interpretation of IR

Interpretation of the NMR and Analysis of the Stereochemistry

In the block below is the chair conformation of a *cis* isomer of the product. In the other block, draw the chair conformation of a *trans* isomer where the alcohol and methyl groups are both equatorial.

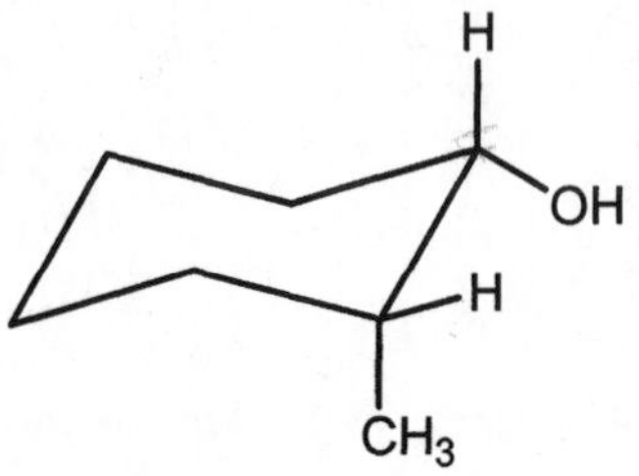

cis

_____ ppm

trans

_____ ppm

Examine the NMR of the product mixture. The signals at 3.1 and 3.8 ppm are due to the proton attached to carbon #1. One of the multiplets is from the *cis* product; the other is from the *trans* product. The number at the bottom of the signal is the integration value which in this experiment provides the ratio of *cis / trans* product. The reason they appear as multiplets is due to spin-spin coupling with neighboring hydrogens. The distance between the peaks in the multiplet is called the coupling constant, or "J" value. It is possible to assign the correct isomer by comparing the proton on carbon #1 to the proton on carbon #2. The "J-value" for protons that are both axial is larger than that of one axial and 1 equatorial, which in turn is larger that that of two equatorial hydrogens. Therefore, the wider pattern will appear for the diaxial protons.

With this in mind, assign the 3.1 and 3.8 signals to the correct stereoisomer and write the ppm values in the blanks below the correct structure above

Now you can decide. What was the major product of the sodium borohydride reduction?

312

EXPERIMENT 38

INFRARED SPECTROSCOPY

INTRODUCTION

In previous experiments you examined several methods such as melting point determination and thin layer chromatography (TLC) for specifically identifying organic compounds. Infrared spectroscopy (IR) is another available method for doing such a characterization. **Spectroscopy** is a technique for analyzing molecules based on differences in how they interact with electromagnetic radiation. A person doing spectroscopy is called a **spectroscopist**. IR identification of organic compounds utilizes the technique of absorption spectroscopy. **Absorption spectroscopy** is the measurement of the quantity of light absorbed by a molecule as a function of frequency or wavelength. Recall from other chemistry and physics courses that the energy of electromagnetic radiation is related to the frequency(ν) of a standard wave by the equation E=hν where "h" is Planck's constant. The standard unit of frequency is the sec^{-1} which is also called a Herz (Hz). This can also be expressed as the wavelength (λ) of the standing wave by dividing frequency into the speed of light (3.0 x 10^8 m/sec). The wavelength in meters can then be converted to centimeters (cm), nanometers (nm), or micrometers (μm). The conventional methods of expressing the type of electromagnetic radiation being used is whatever is convenient for the chemist. For example, people working in the ultraviolet to the visible regions (UV-VIS) of the spectrum express their wavelengths in nanometers.

For years the micrometer (also called the micron) was the preferred unit in IR. The infrared region of the spectrum routinely used in organic chemistry varies from 2.5 μm to 20 μm. More recently the IR spectroscopist has been using the wavenumber to express the energy of IR radiation.

$$\text{Wavenumber (cm}^{-1}) = 10,000 \text{ / } \mu m$$

The infrared region of the electromagnetic spectrum lies between 4000 cm-1 and 500 cm-1. It is in this area of the spectrum where most organic compounds absorb radiation.

The atoms in a molecule are in constant motion called molecular vibrations. Bonds can vibrate with both stretching and bending motions. Stretching refers to a vibration that occurs along the line of a bond. Bending refers to changes in bond angle. Examples can be found on the next page. The vibrations can be symmetric or asymmetric; in-plane or out-of-plane. For organic compounds each of these vibrations occurs at a specific frequency between 4000 and 500 wavenumbers.

The instrument used to determine the absorption spectrum for a compound is called an **infrared spectrophotometer**. A sample is placed in a beam of infrared light, and the frequency is varied from 4000 cm^{-1} to 500 cm^{-1}. At frequencies where specific functional groups undergo stretching or bending vibrations light will be absorbed resulting in a peak A detector generates a plot of the amount of energy absorbed (y-axis) versus the wavenumber of the absorption (x-axis). This produces an IR spectrum consisting of a large number of peaks. The peaks are often referred to as **absorption bands**. IR spectra are rather complex making it rather difficult to look at a final spectrum and determine

STRETCHING AND BENDING VIBRATIONS OF A METHYLENE

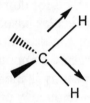

Symmetric Stretching

Asymmetric stretching

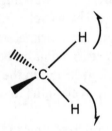

In-plane bending

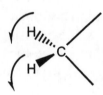

out-of-plane bending

the identity of an unknown compound. However, the stretching frequencies of important functional groups predominate the region between 4000 cm^{-1} and 1000 cm^{-1}. Thus IR can be used to determine the presence or absence of a carbonyl group, an alcohol, or a carbon carbon double bond. A table of the important stretching frequencies can be found on the next page. You will need to use your textbook for a more detailed table. Your text also provides you with samples of IR spectra and their interpretations.

Your lab instructor will provide you with a typical IR spectrum. Before interpreting spectra you need to be familiar with the *organic talk*. **Peaks** (also called **bands**) can be categorized as being "sharp" or "broad." The peaks are also categorized as being very strong(vs), strong(s), medium(m), or weak(w). Along with frequency, these characteristics are often used to identify specific functional groups. The best way of analyzing spectra is to divide the spectrum into zones. You can see this in the handout. APPROXIMATE INFRARED ABSORPTION FREQUENCIES OF A FEW COMMON FUNCTIONAL GROUPS

GROUP	ν (cm^{-1})	INTENSITY
O-H stretch (alcohol)	3600-3200	strong and broad
O-H stretch (acid)	3500-2500	strong and very broad
sp^3 C-H stretch	3000-2800	strong
sp^2 C-H stretch	3100-3000	medium
N-H stretch	3500-3300	medium and broad
C-C triple bond	2300-2100	weak
Carbonyl group	1750-1650	strong and sharp
C-C double bond	1680-1620	medium
C-O ether bond	1200-1050	medium

PROCEDURE

1. On the last page of this experiment is a list of possible liquid unknowns. Your lab instructor will assign you one of these unknowns. They are stored in dropper bottles in the hood. Be sure to write the letter of your unknown liquid in the space provided on the data sheet. Your lab instructor will show you how to operate the Infrared spectrophotometer in the instrument room. Normally the absorption spectrum of a liquid is measured by placing it in a glass cuvette which is then inserted into the sample chamber of the spectrophotometer. However, glass absorbs in the IR region of the spectrum necessitating the use of a different type of sample holder. For IR measurements polished sodium chloride plates (**salt plates**) are used because NaCl does not absorb in the IR. Of course, the salt plates cannot be cleaned with water since NaCl is water soluble. Instead, the plates are cleaned by rinsing with acetone from a squirt bottle. You will find an acetone bottle and waste container in the hood.

2. Place a drop of your unknown on a salt plate. Sandwich it between another salt plate and place it in the sample compartment of the spectrophotometer. Run, mark, and plot the spectrum demonstrated by your lab instructor.

3. Using the frequency assignments you find in your text and lab manual, identify your unknown and complete the data and results page.

4. Your lab instructor will assign you a lab partner who was given a different unknown. Using your partner's spectrum, again identify the unknown and complete the data and results page.

IR Liquid Unknowns

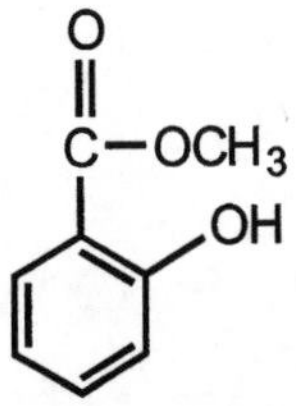

methyl salicylate

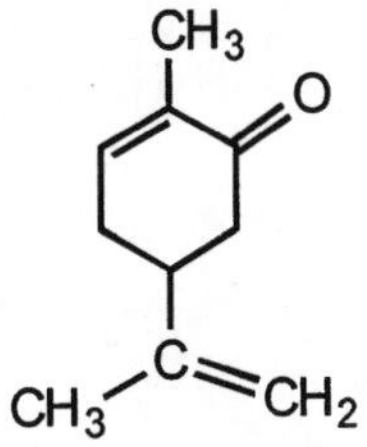

carvone

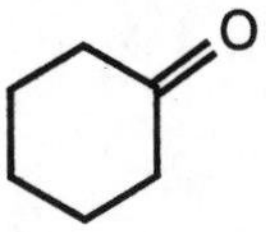

cyclohexanone

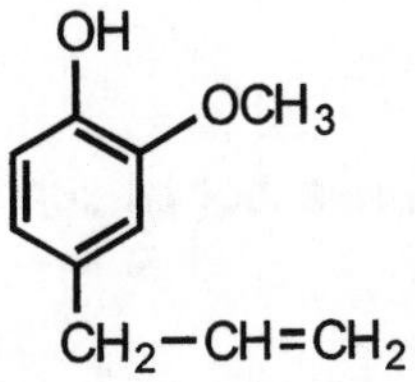

eugenol

$H-C\equiv C-CH_2CH_2CH_2CH_2CH_3$ 1-heptyne

Name _____ Sec. _____

Data and results

Letter of your unknown _____

Identity of your unknown _____

Provide an explanation of how you used the presence or absence of absorption
frequencies to identify the unknown.

Letter of partner's unknown _____

Identity of partner's unknown _____

Provide an explanation of how you used the presence or absence of absorption
frequencies to identify this unknown.

EXPERIMENT 39

EXTRACTION OF CAFFEINE FROM TEA

GAS CHROMATOGRAPHY ANALYSIS

INTRODUCTION

Amines are derivatives of ammonia with one or more "R" groups bonded to the nitrogen atom. They play a crucial role in biological activities. As a result, amines are used in the pharmacological industry in the synthesis of medicines and drugs. One of the well known classes of amine compounds is alkaloids. Since ancient times, they were known as "vegetable alkali" because they were derived from plant sources. In short, amines that are isolated from plants are referred to as alkaloids and contain one or more heterocyclic nitrogen atoms. Nicotine and coniine are well known examples belong to this particular category. In addition, a number of drugs such as morphine, cocaine, and quinine are widely used as painkillers by physicians and belong to this group. These compounds are classified in three major classes as follows:

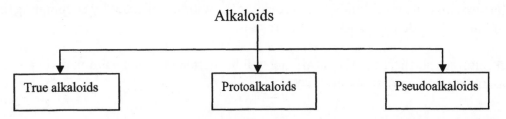

True alkaloids are basic compounds containing nitrogen in a heterocyclic ring. They are derived from amino acids. The second type of alkaloids do not have a nitrogen atom into a ring system. They are simple amines and generally referred as biological amines. Pseudoalkaloids contain two classes: the purines and the steroidal alkaloids. The pseudoalkaloids are not derived from amino acids. One of the well-known and common compounds, caffeine, belongs to the purine type of alkaloid. It contains the nitrogen atom at the seventh position of a pyrimidine ring system that is an essential building block of nucleic acids. Caffeine is a stimulating agent that acts on the cerebral cortex. Therefore, it is used as a major constituent in drugs, especially in alertness aids, such as Vivarin and No-Doz pills. In addition, caffeine is used in headache remedies such as Excedrin.

Tea leaves consists of five main ingredients: cellulose, caffeine, chlorophyll, tannic acids, and flavanoid pigments. The first step the tea leaves are treated with boiled water. This step separates water-soluble material in aqueous tea solution from water insoluble materials such as tea leaves and cellulose. After filtration, the hot aqueous solution contains chlorophyll, caffeine, tannic acids, and pigment.

Caffeine is soluble in methylene chloride; the other substances are not. So the cooled solution is extracted with methylene chloride to separate the caffeine. As a result, the organic layer contains caffeine, and traces of chlorophyll. The crude product constitutes caffeine as a major constituent, and it is later purified by means of sublimation. The aqueous layer contains tannic acid, flavanoid pigment and chlorophyll.

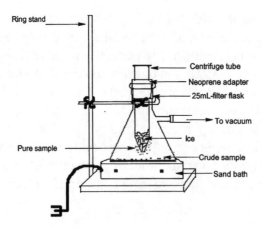

Caffeine   Pyrimidine   Purine

Sodium carbonate is added during the process to increase the water solubility of tannic acids that are already present in tea leaves. It helps to form ionic sodium salts and forms a free base at the end. As mentioned earlier, extraction with methylene chloride will separate caffeine from other substances present in tea leaves. After drying the methylene chloride with calcium chloride, the crude product will be collected by evaporating the methylene chloride with a stream of nitrogen.

Later, further purification of the caffeine will be attained using a sublimation process. The sublimation apparatus can be set up as follows:

To identify the substance, obtain an IR spectrum. To ensure purification of the substance, save some purified sample for a GC (gas chromatography) experiment.

Name _____ Sec. _____ Exp. # _____

Reactant Table

Name of the compound	m.p./b.p.	F.W.	Mass (g)	Vol (ml)	D (g/ml)	mols

Safety/Hazards

PROCEDURE

In a 25-mL Erlenmeyer flask place 1.5g anhydrous sodium carbonate and 12 ml of water. Add a boiling stick and bring this mixture to a boil on a hot plate. After the solid is dissolved, add one tea bag (around 2.4 to 2.5g). Cover the Erlenmeyer flask with a watch glass.

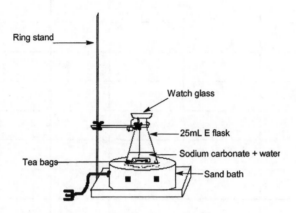

Gently boil this mixture using a sand bath for 10 to15 minutes. Squeeze the tea bag to remove water as much as possible. This can be done by pressing the tea bag gently against the side of the flask. It must be done very carefully without breaking the bag. Cool the flask to room temperature. Meanwhile wash a centrifuge tube and dry thoroughly. Using a Pasteur pipette, transfer the contents of the flask to the clean centrifuge tube. Remove tea bag along with its contents and set it aside. Do not add the tea bag and its contents in the centrifuge tube. Once the centrifuge tube has been cooled to room temperature, the mixture is ready for the extraction process.

Extract the aqueous solution with 2.0 ml of methylene chloride by placing the cap on the centrifuge tube and rocking the centrifuge tube to shake the solution. At this point, extreme shaking will result in loosing caffeine.

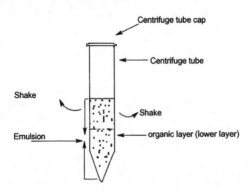

If an emulsion is observed due to some constituents in the tea solution, it can be broken readily by centrifugation. The recommended centrifugation time period is between 60 to 90 seconds. Remove the lower organic layer, and set it aside in the reaction tube. The upper layer emulsion can be left in the centrifuge tube for another extraction. Use another portion of 2.0 ml of methylene chloride and extract one more time. Remove this organic layer and add it to the previous saved organic layer in the reaction tube.

Add anhydrous calcium chloride pellets to the reaction tube and dry the combined extracts for 5- to 10 minutes maximum. For the best result, add drying agent in small portions and shake the solution. Add until no clumping is observed. Keep solution undisturbed for 5-10minutes. Meanwhile, weigh a 25-ml filter flask and record its weight. Decant dried solution into this weighed flask.

Rinse the calcium chloride pellets with the addition 2.0 ml of methylene chloride. Do this three or four more times. Collect these washings in a 10-mL Erlenmyer flask and dry with the anhydrous calcium chloride. Transfer the dry mixture to the 25-ml flask. Evaporate the solvent under nitrogen flow or air supply. If these are not available readily, add a boiling stone to the flask and evaporate the solvent by warming the flask in a sand bath in the hood. The crude caffeine product will be formed as an off-white crystalline solid with a green tinge. Cool and weigh this flask. Save this crude product for purification and TLC.

Begin the purification of the caffeine by assembling the sublimation apparatus shown earlier. Clean a centrifuge tube and fill it half way with ice cubes. Fit a neoprene adapter on top of the filter flask. Insert a clean centrifuge tube through this neoprene adapter. Secure the filter flask using long utility clamp to the ring stand. Connect the side arm of the filter flask to vacuum using rubber tubing. Once the apparatus is set up, apply a vacuum to the system through the filter flask to evacuate the system. Start heating the filter flask gently using a sand bath. Rotate the filter flask as heating is continued. Pure sample will start depositing on the surface of the centrifuge tube. Remove the heat when no more caffeine will sublime. Shut off the aspirator and allow the apparatus to cool to room temperature. Remove the ice from the centrifuge tube. Do not remove centrifuge tube until the flask is cooled down.

Scrape the pure product onto weighing paper. Weigh the purified product. Save pure caffeine for IR and TLC purposes.

Run a TLC of crude and purified product using 95% ethyl acetate and 5% acetic acid as the mobile phase. Calculate R_f value and record it in the data table. Turn in the TLC plate with the lab report.

To run a GC save the pure sample in a labeled vial for the next week's experiment. If time permits, run a proton NMR of the pure product and analyze the spectrum.

Name _____ Sec. ____ Exp. #____

PROCEDURE	OBSERVATIONS

Name _____ Sec. _____ Exp. #_____

PROCEDURE	OBSERVATIONS

Name _____ Sec. _____ Exp. # _____

DATA AND RESULTS

Weight of the product:

TLC

Calculation of R_f value

Paste TLC plate below:

QUESTIONS

1.   Why does the crude product have a slightly greenish in color?

2. Why is it called isolation of caffeine rather calling extraction of caffeine from tea leaves?

3. Draw a flow diagram for isolation of caffeine from tea leaves.